W9-BPR-639

Here's what they're saying about
The Art & Science of Dumpster Diving

"For a how-to book, this is awfully entertaining. Thought provoking and conscience-stirring, too. Amazing and amusing..."
— **Booklist (Recommended)**

"*The Art & Science of Dumpster Diving* does not lack either an edge or an attitude. What a hoot of a manual... I can guarantee Hoffman's manual is unlike anything you have ever read. Perhaps it's right up your alley."
— **The Indianapolis News**

"152 pages of everything you need to know about stalking the wild dumpster but were afraid to ask."
— **New Reality**

"With its helpful diagrams and photographs, its forays into history and love, its advice on topics as diverse as hygiene and law and cuisine, and its always loving, careful attention to fine points of technique, *The Art & Science of Dumpster Diving* is sometimes bizarrely reminiscent of the *Boy Scout Handbook*."
— **USA Today**

"*The Art & Science of Dumpster Diving* [is] a practical guide to rooting around in other people's trash and garbage, everything from the proper clothes and equipment to the proper response when a suspicious cop or security guard jacks you up."
— **New York Press**

The Art & Science of
Dumpster Diving

by John Hoffman

with an Introduction by Jim Broadstreet,
Author of *Building With Junk*

and Original Comix byAce Backwords,
Creator of *Twisted Image*

Loompanics Unlimited
Port Townsend, Washington

The Art & Science of Dumpster Diving
© 1993 by John Hoffman

Introduction © 1993 by Jim Broadstreet

Comix © 1993 by Ace Backwords

Published by:
Loompanics Unlimited
P.O. Box 1197
Port Townsend, WA 98368
U.S.A.

Loompanics Unlimited is a division of Loompanics Enterprises, Inc.

Cover by Ace Backwords, Colored by Barbara Williams
Illustrations by Kevin Martin
Photos courtesy of John and Tina Hoffman

ISBN 1-55950-088-3
Library of Congress Catalog Card Number 92-074645

Contents

About the Author

John Hoffman has been a dumpster diver since childhood. He is a third generation dumpster diver.

In addition, he has worked as a psychiatric behavior counselor, a printer's assistant, a security officer, a pizza cook/delivery person, live bait salesman, newspaper editor/reporter, a historical research assistant, a hotel clerk, and has done a bit of poaching... among other things.

His formal education includes a degree in English Writing, *magna cum laude,* from a private Lutheran college in Minnesota. He has written numerous newspaper articles, opinion columns, poems, short stories, and has contributed to local history books.

He recently embarked on a "five year plan to self-sustaining wealth." He is also writing a science fiction trilogy which emphasizes personal evolution and liberty.

A Note from the Publisher

We found this manuscript in the trash. Honestly. It was at the bottom of a very tall pile of unsolicited manuscripts. We were slashing through the pile when we opened this up, looked at the cover letter and said, "Who's going to buy a book on dumpster diving?" Since it wasn't accompanied by return postage, we threw the manuscript in the trash.

Then we saw the photos. Pictures of a guy surrounded with grocery-store abundance. Pictures of a guy with flowers from the dumpster. Picture after picture of piles of garbage-picked wealth and trashy entertainment. It was amazing and, best of all, it was *real.*

We pulled the manuscript out of the trash and read the first chapter. We were hooked. You will be, too.

Introduction
by Jim Broadstreet
author of *Building With Junk*

Have you noticed a change taking place in America lately? Not just that we have become the most indebted people in history but all the other stuff that goes with it?

Does it ever occur to you that your children, and theirs are not going to enjoy everything that we have? Indeed, they are going to be paying the tab which we have run up, and which we are adding to, continually and with reckless abandon, as if we don't give a damn — as if there will be no tomorrow.

If you, the reader, were "born with a silver spoon in your mouth" you have been lucky. If you were and you assume that you will hold on to it — that things will just naturally remain easy and comfortable, with minimal effort on your part, then you are living with a terribly false assumption — one that could lead to horrendous disillusionment, even personal disaster!

We no longer populate a "land of milk and honey" from which all of our needs and desires will be automatically fulfilled, regardless of effort and thought expended.

The "survival of the fittest" concept of life almost died with the spiraling affluence which spread across this country after World War II. That spiral has topped out. Now, as we slide out of our position of global pre-eminence, those of us with cunning, ingenuity and some daring are going to "survive" better than the rest.

"Almost" died, I said above. In this land of enormous diversity there are always exceptions. We who are fortunate enough to have come upon this book, *The Art & Science of Dumpster Diving*, have the advantage of being able to learn from someone who is, along with his wonderfully different family, an exception from the norm — big time!

John Hoffman's family, as you are about to learn, were almost destined, by virtue of circumstances well beyond their control, to be recipients of welfare. With welfare would probably come low self-esteem and the myriad problems associated with it. Not only did that not happen but every member became, not just "productive," but super productive!

Not only did the Hoffmans "carry their own weight" in their community, they were leaders in it. They were, and still are avant-garde in environmental action. And now John Hoffman has given us this gift — this compilation of knowledge assembled for us to use to our greatest benefit. All we need do is cast off some of our societally implanted and destructive codes of behavior to allow us to become members of the "chosen tribe." Not only will we be ahead of the pack in adjusting to the coming new American way-of-life, but we will be leaders in changing society's attitudes concerning consumption, waste, negligence, and sloth.

Now, if the above words have led you to assume that this is a book of heavy reading — one that would have to be struggled through for the sake of learning what you may, or may not, be able to use — well, not so!

John Hoffman has seen a lot in his day! His experiences have been, as with all of us, every thing from mean and ugly to joyous and fun. He has the ability to see the funny in almost everything and, fortunately for us, to pass that humor on in his written words.

Then too, Hoffman and his family, friends and cohorts have developed what you might find to be some rather bizarre philosophies-of-life! It is certainly not necessary to agree with all of these but it is a lot of fun to see how some of them developed and to be fascinated by the tangents taken by this creative mind which was, seemingly, never discouraged from pursuing any avenue.

Reading *The Art & Science of Dumpster Diving* will provoke you to laugh a lot, cringe a little, feel some sorrow for our society, get a little angry and, perhaps, change some of your deeply instilled concepts of what life is all about — and how it might be lived a whole lot more sensibly.

TWISTED IMAGE by Ace Backwords ©1993

Chapter 1
Dumpster Dinner

Some weeks ago, I finished a nice hot shower which happened to include dumpster-salvaged soap, dumpster-salvaged shampoo, a dumpster salvaged towel and a dumpster-salvaged bath mat.

I slapped on my wife's favorite cologne, which came from... yes, a *dumpster*. And then, naturally, I shaved. My job at the hospital requires meticulous grooming and cleanliness. Now, it's true I purchased the razors, but the shaving cream came from a dumpster. I also purchased the deodorant, but in the past I've salvaged deodorant from trash bins.

That's right, *deodorant* from trash bins.

Checking my watch (a freebie from the dumpster) I hurried into the kitchen and sat down at the table. The dinner table was also a freebie, as well as the chairs. We sold the table we had before *this one* for $25 and paid the electric bill with the money.

I noted that my wife had her "food science" textbook on the kitchen counter. She has a degree in Biology, but "food science" was an elective she studied at our private college. I also have a college degree.

"Something special?" I asked.

"Wait and see!" she smiled.

While waiting, I watered our half-dozen or so plants. All but one came from dumpsters. I sat at the table again, and my wife set a steaming plate down in front of me.

I noted that she was wearing the 1920s sterling silver "butterfly" pin that I had dumpster dived a week earlier... only two hundred yards from our apartment, *swear to God*. And that wasn't all I found that time.

"Looks great, honey!" I said, meaning the food.

My wife, Tina, had prepared steamed artichokes, vegetable-beef stir fry, rice, and fresh-baked bread from instant dough. For dessert, an assortment of fruit. Only the beef in the stir-fry did not originate in the dumpster. Oh, yes, and the soy sauce. My wife always grabs several packets of soy sauce whenever we have the "egg roll special" at the local mall. The dinner plates were a wedding present, but the silverware came from a dumpster.

"What do you want to drink?" Tina asked. "Milk? Orange juice? Iced tea?"

"Just water, " I answered. "All this food is making me so *fat*."

Both the milk and the orange juice originated in the dumpster. The tea, however, came from the hospital cafeteria. I never eat there, but whenever I escort a patient, I grab several bags of tea.

"Don't forget to take those magazines to the hospital," my wife said.

"I'm still reading the *Rolling Stone,*" I answered.

"That reminds me," Tina said.

Hopping off her chair, she turned on the television (from a dumpster) and turned the channel to MTV. At the moment, MTV was featuring a music video from a very popular (and *rich*) singer/movie star. I won't mention her name, but let's just say she often causes controversies over "censorship versus art."

"You know," I told my wife, "*she* used to be a dumpster diver."

"I know she posed nude," my wife replied, "before she hit the big time."

"She ate stuff from dumpsters, too," I said.

Ironically, I had read that in a magazine obtained from a dumpster. This *certain* pop star, while still an anonymous face in the crowd, had posed nude and scavenged food from trash cans. Now, obviously, she didn't proceed to launch a multi-million dollar career in recovering foodstuffs and recycling soda cans. Using her brains and talent, *doing whatever she had to in order to survive,* this young woman never wavered in the pursuit of her dreams and today she is a millionaire pop star.

Yes, her *talent* and her *brains* made her a star, but dumpster diving gave her a vital Darwinian advantage. How many other young, talented men and women have shared apartments, worked in hamburger joints, pounded the streets day and night in pursuit of their elusive life's dream? Instead of success, most encounter only closed doors, rip-offs, poverty and hunger. As long as possible, they struggle for their dreams, they keep believing in themselves.

Why? Why do people fight and fight and *fight* for their dreams? It's a rhetorical question. You know the answer. *Because it's your dream.* It's your *life.* The universe as you know it revolves around your eyeballs, the things in life that matter are the things that matter to *you.* You have, so far as we know for certain, one life. You have one opportunity to do, to be, to experience, to create.

However, you're in competition with a lot of other highly-motivated individuals. Like you, these people are intensely goal-oriented, utterly devoted to their own dreams, careers, or families. The world is like the jungle, the woods, the sea... in our society's "economic ecosystem," very little drops to the ground or washes up on shore without becoming food or fertilizer for *something.* And that *something* is right there, waiting, mouth open.

Look around you — try seeing *economic* activity in terms of *nature.* For every opportunity, every windfall, every resource, every niche... something is already sitting there, making a living, getting a cut, earning interest, drawing a percentage, running the action. Only the leanest, the hungriest, the smartest, the most motivated and calculating and utterly devoted will achieve their dreams. Only the most able *deserve* their dreams.

It has been said that "all the world is a stage." Well, most people never get beyond the cattle call... or the casting couch. Most of the time when the world offers you a "big break" it is, in fact, offering to screw you and never return your phone calls.

You need an edge...

Consider that hungry, highly-motivated pop-star-to-be. Do you imagine that while seeking stardom she lived under bridges and in homeless shelters... slept in a cardboard box... wore a baggy overcoat and Army boots? That she smelled bad? That she washed her hair in public restrooms? That she ate out of a dumpster?

Dumpster diving is *no longer the action of last resort.* Dumpster diving, in fact, can be your edge, your vital Darwinian advantage.

Dumpster diving — an activity pioneered by bag ladies and homeless ex-mental patients — is becoming more and more practical and *profitable.* So profitable, in fact, that it can make the vital difference between attaining your dreams.... or returning home on a borrowed bus ticket to work at Daddy's hamburger stand.

Dumpster diving does *not* mean scavenging amid somebody's kitchen scraps, consuming half-rotten, half-eaten chicken legs ala Hefty Bag. Yes, *some* people do that — and those people need a hot shower and mental health care, badly. Those people commune regularly with the Space Brothers.

Your modern dumpster diver, in contrast, may be a full-time student, an apartment dweller, a semi-rural seeker of self-sufficiency, or a young, educated professional — like myself. A modern dumpster diver may be somebody who chooses to work *less* and spend more time in pursuit of dreams, goals, activism, *art* — like that pop star.

Which brings us to an explanation of our title, *The Art and Science of Dumpster Diving.*

What is "art?" Obviously, I can't answer a question in a couple of paragraphs when the question can't *begin* to be answered in volumes of books and hours of discussion. But consider this: painting pictures is, without a doubt, *art.* But so is photography, architecture, and pottery.

So, it turns out, something can be *functional* and be an expression of creative self.

Now, I've heard medicine called "the healing art." I've also heard cooking, writing, and wine making called arts. These things are often more functional than expressive, but are still considered "arts."

Art is quite a broad topic, no? Think about this: once an artist, *always* an artist. What does that mean? It means this: a writer is not only a writer when he *writes.* You are an artist, a *creative being,* twenty-four hours a day.

When you cook, when you drive, when you speak, when you labor, when you do whatever it is you *do,* you are still expressing yourself, pioneering ideas, concepts, movements, *style.* Whose creative goals do you want to triumph? Your goals, of course. Whose dream is important? Your dream. Why? Because it's YOUR life and the whole universe is at your fingertips for YOU to experience, for YOU to change as YOU see fit. After all, you're an *artiste.*

You deserve to triumph. You deserve your dream because you *want* it, because you refuse to live a life of quiet desperation, because you're lean and hungry and right there on the cutting edge of Darwinism. YOU are tomorrow's dominant life form. You *will* attain your goals, because you're willing to fight, willing to do anything within the bounds of YOUR rules, YOUR reason — even pick up the book called *The Art & Science of Dumpster Diving.*

Good move.

And, I might add, you're in good company.

If you have preconceived notions handed to you by society, suspend those notions for a little while. The lid is lifting, you're about to enter a whole new universe as you learn *who* dumpster dives and *what* they discover.

Welcome, diving comrade. Welcome, to *the art and science...*

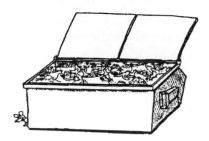

Chapter 2
A Realistic Path to Self-Sufficiency

Ah... memories. Most individuals carry around pleasant, idealized remembrances of the fun they had as children. I probably have more of these golden tinted memories than most people, since I spent so much of my childhood digging around in dumpsters.

Yes, digging around in dumpsters. Oh, to be six years old again and feel the thrill of finding a big ol' box filled with my favorite snack food... or a bike, needing only minor repairs... or even mundane items like a stylish jacket for school wear. As an adult dumpster diver I have located food, furniture, even valuable antiques but very little matches the sensation of "Christmas every week" that I felt as a youngster.

Perhaps you think I'm kidding. Maybe you think I had a terrible, impoverished childhood and what I'm saying has a sarcastic, bitter undertone. I'm not joking, however. I had a wonderful childhood, wise and kind parents, and the family activity I enjoyed the most (besides fishing) was dumpster diving.

Thanks to this pursuit, our family lived happily with abundance while families in similar circumstances barely scraped by and destroyed their own flesh with fighting, screaming, drinking and worse. I will always be thankful that my parents were pioneering masters of the fine art. For decades, members of my family kept the knowledge contained in this book a closely guarded secret. We protected this secret more carefully than some families protect ancient recipes and knowledge of horrible crimes.

And it was because of our secret source of abundance that I can look back and say, "ah, memories" while most of the kids on my rural school bus route experienced misery and now have miserable families of their own.

Poverty. *Miserable* poverty. *Humiliating* lack of material goods. Our neighbors had all that and much, much more. Supposedly alcoholism, incest, and physical/emotional abuse can strike families across the socioeconomic spectrum. However, when a family is blessed with abundant resources it is not likely that the children will go to school unwashed and hungry... even if the father is an alcoholic. Not so with poor families. Let's face it: material resources can take the sting out of a hell of a lot of weakness and vice. But our neighbors didn't have material wealth — far from it.

The Bembenek kids... the Kietzer twins... the whole Ruben clan. No, these are not real sur-names. I will not take the risk of humiliating the actual families. But humiliation was a daily experience for their children.

They were *always* late for the bus. They'd come dashing out of their poorly painted, run down houses when the bus driver leaned on the horn. Their clothes were unwashed, tattered, and *familiar* — the same clothes they wore all week. Their lunch — if they had one — was hardly enough to satisfy a small dog. These kids were *gaunt*. They smelled bad. Their hair was unwashed, uncombed and they had poor complexions. They had plenty of broken bones, sprains, stitches and periods of prolonged "illness." Needless to say, these kids did poorly in school. They didn't have the self esteem, concentration or energy to get decent grades.

These were the children of the "rural poor." Their parents had measly parcels of land which they tried, halfheartedly, to plant with crops or use for livestock. Most of their income, however, came from jobs in the nearby town. And, on paper, my family had no more wealth than our neighbors. In fact, we appeared to have less on paper.

We had about five acres of land with small farm buildings constructed in the 1930s. My father and mother married when he was fifty-eight and she was thirty-eight and promptly proceeded to have three children. Our "visible income" consisted of veteran's disability checks, social security, and whatever government giveaways one could obtain by filling out a form at the local courthouse. From time to time my father and mother would work at a wide variety of jobs if we wanted cash for a major purchase.

Many of our neighbors had more land and/or more formal education, and did not carry around a souvenir of Leyte Gulf in their spinal column. However, these other families were constantly distressed about "making ends meet." We had all our ends met and a surplus, besides. Why? You know why... *the fine art...*

Maximum Diving Lifestyle

For approximately two decades, my family lived what I will describe as a "maximum diving lifestyle." This does *not* mean we obtained all our income and possessions from dumpsters. Rather, what I mean by this is that we enjoyed the most benefits that can, in all likelihood, be derived from dumpster diving. If we had not enjoyed so many "perks," if we were only now and then dumpster dippers, I would never consider writing a book telling others how to gain by dumpster salvage. However, as it turns out, we enjoyed a lifestyle that can only be described as "idyllic" while families in similar circumstances damn near perished. The only difference between us and them was a little bit of common sense, a little more innovation, and a lot of good stuff that we salvaged from trash bins.

So why am I telling YOU? Why not keep all the goodies for myself?

As I stated, we did exactly that for a long time. Through dumpster diving. "Willard and Vernie" Hoffman managed to abundantly provide for three children. Now all of us kids are grown and have productive, happy lives. Willard has gone to his reward and my mother is in comfortable circumstances. Dumpster diving is still an important part of my life and the lives of my brother and sister. But, thanks to our parents and our own motivation, we have obtained valuable, moneymaking skills. Our schoolmates, in contrast, are even more miserable than their parents. Some are in jail, mental institutions, or dead.

I'm not doing this out of pity, however. I'm writing this book to line my greedy pockets. And I'm going to spill the beans, all the beans, so I can make as much money as possible. It's time to take the "secret recipe" to market and start selling the sauce.

What's so great about the fine art? Let me tell you about those "golden memories."

F-O-O-D. My god, did we have food. We had two huge freezers stuffed with frozen yummies. We had a root cellar filled to bursting with canned goods and produce. We had cupboards and closets and a fridge stuffed with chow. Sometimes Dad would open a closet and jump back to avoid a cascade of cans. We had emergency stashes of goods in the attic, the barn, and carefully buried in the yard, "just in case."

Outside, chickens ran around — laying eggs, cackling, awaiting their fate as "chicken ala Vernie" or soup. Or both. Waste not, want not. Sometimes we had a pig, a goat, ducks, geese or other livestock. We favored chickens, however. Less work and fun to watch. Animals are a lot of responsibility and we often preferred to spend our time fishing, hunting, reading, or engaged in productive personal hobbies.

We had a huge garden. Orchards. Herb planters. A strawberry patch. Raspberry bushes. Gooseberries. For a few years we had hundreds of rabbits, but then we went back to chickens. We tried mushrooms, too. Every year we tried some of the "unusual" produce advertised in some corner of the seed catalogs. Experimenting with different plants and livestock was something we had the time and ability to pursue.

Those of you who are interested in "self-sufficiency" but think it is out of reach — keep in mind that we managed all this on a "visible income" that most people couldn't use to provide for *two* people, let alone two older adults, three teenagers, assorted cats, dogs, parakeets, tropical fish and livestock. Dumpster diving was the secret.

We had so much food we traded it for stuff. Even in unexpected circumstances we never lacked food and, frequently, made deals with neighbors that saved *other* families from desperate times of need... and provided us with their skills or certain goods. Our biggest problem with food was concealing the fact that we had so

much. This was particularly a problem when our "visible income" qualified us for food stamps, which we accepted and used. Remember, when something good is being thrown out or given away *grab as much as you possibly can*. That was how I felt about the thousands of dollars I received as an "impoverished" student striving toward a college degree.

Ah, memories. We had birthday parties that were the envy of the whole neighborhood. Each of the Hoffman kids had a bike, and frequently we had several bikes until we made trades. We had a respectable stash of firearms and ammunition. We had more good, decent clothes than we could wear. Once, as a first grader, I told my mom that the new teacher's aide had asked me if I had "a pretty scarf for every day of the month." It turned out that I did, indeed, have more than thirty scarves.

And all the Hoffman kids excelled in school. This was partially because of our parents' help and encouragement, but I believe a great deal of it was due to the fact that we had several thousand books in the "library room," hundreds of various record albums, and a number of musical instruments. We had an endless supply of drawing paper and materials for art and science projects. We had the time and means to pursue various extracurricular activities.

And, throughout everything, my parents engaged in various forms of activism and fought battles of principle against organized idiocy in all its many forms.

Some neighbors thought God had blessed us because my mother was devoted to Him. Others thought my father, a veteran of the Pacific Island hopping campaign and the Pearl Harbor attack, had some kind of incredible survivalist secret. The worst of our neighbors — the shrieking, hysterical, pity-seeking neuro-bitches — thought it was simply because the Hoffmans had three kids who were smart, hardworking, and "didn't break their mother's heart like worthless, no-good assholes!"

The "Big Secret," however, was that Mom and Dad, and the three kids spent a little bit of time every day digging around in dumpsters. If God blessed our Mom, it was with an ability to see the value in discarded items. Dad's "survivalist" abilities consisted not only of skillful poaching, but to a greater extent, skillful dumpster salvage. And, as much as we kids excelled in academics, there was one honor for which we only competed with each other... and that competition was fierce.

We all wanted to be "Master Diver."

Abundance And Self-Sufficiency

As a child, I never realized that our lifestyle was incredible. It just seemed natural that we had abundance and happiness while other families did not, despite similar income and circumstances. If I ever asked a question about our good fortune, I was informed that the Hoffmans were simply a little more careful and didn't waste so much. It seemed reasonable to me. And I was told again that I must never, ever tell *anyone* about our dumpster diving. If we told, the other families would simply get there first and grab all the goodies. We didn't even use the expression "dumpster diving" but called it "looking behind Jerry's." Even when the school bus bullies taunted us that the Hoffmans weren't so smart, after all, we never told The Secret. Once, my little brother, Jedediah smiled and said, "Oh, yeah? Well, I've got a secret."

"What is it?" asked a big, smelly member of the Bembenek litter.

At that moment I and my sister, Rebekka, gave little Jed a look that made him turn pale.

"Nuh-nuthin." Jed stammered. "I guess I don't have a s-s-secret, after all."

The amazing thing was that plenty of people actually knew we scavenged dumpster materials. But if anyone asked what we were doing in the alley behind Jerry's Food Mart, my mother would remark casually that she "needed a few boxes" and change the subject. Nobody knew

that we were tapping an incredible source of wealth and we damn sure didn't tell them.

If YOU happen to be one of the "rural poor" or a person *dreaming* of a little piece of land and self-sufficiency, *boy, is this the book for you!* However, if you are an apartment dweller, a student, a struggling homeowner, an activist, a radical, a starving artist or even a "modern nomad" in a van or RV, this book can teach you vital skills that can save your ass — and your dream.

Diving Is Believing

"If the stuff is so good, why did somebody throw it away?"

This is a question I encountered many times when I first began to share my experiences. And, years later, I still don't have an answer. I've provided facts, figures, and my own anecdotal experience to show the wealth is, in fact, there. *Why* is another matter.

Why has most of America forgotten the bitter lessons of the 1930s?

Why do people experience unemployment and malnutrition while surrounded with opportunity and food sources?

Why do people spend money they don't have to buy things they don't need? *Why* does the government print money to give away to Third World ratholes while oppressing its own citizens with productivity-hindering taxes, laws and regulations?

I don't know. But I can tell you this: THE UNITED STATES IS FULL OF IDIOTS DISCARDING PERFECTLY GOOD MATERIAL WEALTH.

It's a diver's market. Don't ask why, just dive, baby, *dive.*

Dumpster diving is like having a generous uncle with lots of stuff. You drop by your uncle's place and say, "Hey, Unc... nice chair you have there."

"You want it?" Uncle replies. "Take it off my hands."

"Thanks, Unc!" you reply, and toss it on the back of the truck.

"How about twenty frozen gourmet entrees and a brass planter?" Uncle inquires, politely.

"Sure!" you reply. "I'll eat the grub and sell that brass planter."

"Well, as long as you're selling stuff," Uncle says, thoughtfully, "I have, somewhere here... a crystal candy dish and a whole box full of perfectly good videotapes. And a small end table. By the way, I've got a nice shirt that would fit you. Oh, how about all these old picture frames?"

"Thanks, Unc!" you exclaim. "I'll drop the stuff off at the consignment store. I'm sure the money will buy me a couple of tanks of gas. Now I can afford to take my girl out this weekend."

"Hey," Uncle says, with a wink. "I've got three dozen *Playboys* from the 1970s packed in acid proof plastic — in sequential order."

"Whoa!" you exclaim. "You're the greatest, Uncle!"

"Think nothing of it!" he replies. "And please, come by tomorrow and take a few more things off my hands."

Everybody should have a kind and generous uncle, don't you think? Better yet, YOU should have one. Everything I named in that little vignette was stuff I've actually found in dumpsters... this month. I've made tens of thousands of dives in a variety of locations and circumstances, but every week I find something that intrigues and amazes me. I'm convinced that, sooner or later, just about anything will be thrown out.

Whatever your concern, whatever your interest, somebody is probably tossing out something you can use right this minute.

Food. Lots and lots of good, usable food. Clothing. Tools. Building materials. Every kind of household furnishing. Toys. Repairable and working appliances. Craft and hobby materials. Sports equipment. Books and magazines... by the ton. Valuable scrap metal. Live plants, planters and all. Informative documents and papers. Stuff suitable to feed livestock. Composting material and things suitable for fertilizer by the ton, by the truckload, mountains of it, neatly bagged for transport.

And much, much more. Scary stuff. Firearms. Human bodies. Live babies. Drugs. Money. Yes, *money.* Identification. Credit cards. Uncashed checks. Blank checks. I found five books of blank checks just the other day, within a block of my residence. Letters suitable for blackmail. Pornography. Pictures of your neighbors having sex. "Survivalist" books... like this one. It's hard for me to believe, and I've been doing this for years.

The other day I found, beneath a pile of personal papers, a bag with over a hundred coins in it. Most of the coins were relatively unimportant pieces from European countries. However, there were two Kennedy halves and a silver "liberty head" half dollar, condition "very good," date 1914, mint mark "S." I *frequently* find small amounts of change, unused postage stamps, and books of blank checks. I pull in a charge card or a book of checks at least once a week. Naturally, I don't use these things for fraud, but the potential is there.

Everything produced by our society is eventually discarded by the same super wasteful society. To think that in the 1930s and 1940s we were boiling bones twice to make soup. Today, within the lifetime of that same generation, we are discarding valuable property as though every household had its own magic lantern...

This situation can not, *will not* continue forever. But, today, you can cash in. Do you want to achieve self-sufficiency? Freedom from want? Freedom from coercion? Do you want to aim your resources at your *goals*, and not squander your financial firepower on day-to-day needs?

Dive, brother.... and save your dimes for a rainy day.

The Hard Realities

Friend, my intention is not to "b.s." you or pull the wool over your eyes. When I say the wealth is there, *it's there*. However, if I have conveyed the impression that dumpster diving is like going to a yard sale or picking up several pricey items for free, or stopping by a supermarket and loading your cart with food — well, that's incorrect.

Gain requires some smarts and some effort. However, considering the gain involved, dumpster diving requires less physical effort and less cerebral expenditure than most activities. And, this time, I can tell you why.

1. As mentioned previously, THE UNITED STATES IS FULL OF IDIOTS DISCARDING PERFECTLY GOOD MATERIAL WEALTH.

2. The United States is full of uptight individuals who wouldn't dare poke around in dumpsters because of vague fears about "germs," "laws," "vermin," "socially unacceptable behavior," and other self-erected mental prisons.

Done correctly, dumpster diving does not have to be messy, embarrassing, dangerous or time consuming. But it *does* require some willingness to go against the norm.

Besides all the dumpster goodies — our focus of concern, here — there's a lot of other crap you don't want. The funny thing is that, many times, what you don't want today you may want next year, or somebody else may want it. To a great extent, this book takes that into account and tells you how to store certain items, make trades, etc.

However, at all times I am acutely conscious of the effort versus benefits ratio. Damn near everything has value to somebody at some time. That doesn't mean, however, that it's always worth your time to haul off a load of lumber scrap — particularly if you live in a small apartment and don't know anybody nearby with a wood-burning stove. That's what's so great about being a semi-rural dumpster diver. Finding materials useful on a farmstead is about as hard as falling off a log.

It's heartbreaking for me to live in an apartment, drawing my wealth from a career in the city, watching people discard perfectly good lumber, cinder blocks, firewood, sheet metal, glass, insulation, plastic piping, ferrous items, animal feed, compost and fertilizer material. The only thing worse is to watch people strive for "self-sufficiency" by spending their hard-earned money on commercially manufactured utility buildings, greenhouses, planting boxes, animal feed and fertilizer.

Rural living is great, but it's a lot of work, too. Animals have to be fed, watered and kept in good health, gardens tended, buildings kept in good repair, firewood cut and stacked. Digging up potatoes and butchering pigs is hard work.

There are plenty of advantages to urban living, advantages that can be increased with dumpster diving. I don't have to mess around sawing wood, stacking it, hauling it around, lighting fires and keeping the fire going. Instead of throwing all the firewood in the truck, I can simply poke through it until I find — let's see — a perfectly good bookshelf. I can turn up the thermostat, sit back, admire my new bookshelf and read up on a favorite topic. Urban or rural, when you feel hungry from reading or chopping wood, you can walk to the fridge and pull out one of those frozen gourmet entrees.

Yes, everything has potential use and value to somebody, somewhere. That doesn't mean you should save every cardboard box, plastic bottle, and glass jar. And that's what dumpsters are full of, frankly — packaging.

Books, shelf and tape case (with tapes) found in dumpster.

Changes In Society

Tons and tons of it, diving comrades. Dumpsters aren't full of stinky refuse, dumpsters are full of *packaging*. Boxes, bottles, wrappers, jars, bags and tubes and you name it. Everything purchased in a store is wrapped and boxed once, twice, maybe three times and put in a disposable grocery bag, too. Think of all the time you've wasted in your life just *unwrapping* crap.

Once upon a time people produced or traded for most of the items they consumed. If they purchased something in a bottle, crock, or can they somehow used that container. I've seen beautiful religious portraits produced from old oyster tins, hammered flat, engraved and painted. At old farmsites I've noted "Prince Albert" cans used to patch buildings, newspapers layered into walls for insulation, boards from shipping crates hammered into chicken coops and corncribs. Even in large cities, one hundred years ago, "ragmen" collected and sold household discards.

But that old way of doing things fell by the wayside. People became more and more detached from producing products for their own use and consumption. Pretty soon even farmers — once self-sufficient — became dangerously specialized and relied more and more on mass produced items.

Now, I'm not one of those weepy basket-weaving types who pine for the day when everyone had to pick their own berries and sew together their own clothes. How would you like to be back in the "good ol' days" with polio and chamber pots?

However, there's a flipside to the present state of affairs — massive waste. People have become so specialized, so detached from the fabric of society, so dependent on their own little niche, so caught up in their own abstractions, they can no longer see the value in an old piece of furniture, slightly bruised produce or a pile of lumber.

Maybe a highly-paid lawyer/woodtick-on-the-ass-of-society can afford to toss out his old books, furniture, wardrobe and videotapes. Maybe stores can afford to discard tons of usable food. Yes, there's something wrong with all that waste. But that's the way things are — why shouldn't you benefit?

Profit From Stupidity

Digging in other people's garbage.

Take a good look at these words and think about your gut reaction. My eager comrade, dumpster diving is one of the great American taboos. It's so taboo there aren't even organized groups fighting against it — yet. Once I called a live radio talk show and shared a little bit about dumpster diving grocery stores. The host hardly spoke a word — he just let me say all these shocking, controversial things. I shared some of my philosophy with the radioland audience, stuff like "just because a box of food has seen the confines of a dumpster *does not* render it unfit for human consumption. The inside of a dumpster is about as sanitary as the inside of a produce truck — often more so."

Call-in reactions were immediate and livid. One particularly stupid individual wondered aloud if I would sue the store if "tainted" food made me ill. Probably one of those lawyer/-woodtick types. Another idealistic dim wit said hungry individuals should go to the "local food shelf," not dumpsters. Who said I was hungry? And who said I don't already use freebie food programs as often as I can get away with it?

The pro-waste mentality is a part of the very fabric of our nation's true belief system. The wealth is there, and *everybody is afraid to grab it.* Afraid to look "poor." Afraid of "getting into trouble." Afraid of "invading somebody's privacy." Or just ignorant of the possibilities.

And that's precisely why the opportunities are so great for the daring few. People like YOU.

Good Days And GREAT Days

I seldom if ever return home empty handed after dumpster diving. Very few things in life give you a freebie each and every time you show up. Those rare times when I come home empty-handed can usually be traced to one of the following reasons.

1. I just didn't look very hard or very long — maybe I had other things to do that day and only checked a few dumpsters on my "trapline."

2. Somebody else was there first.

When my dad and I returned from fishing empty-handed, my dad called it "getting skunked." All the same, it was fun to be out fishing, netting or spearing. Dumpster diving will leave you empty-handed more rarely than fishing, trapping or gathering wild edibles. But, like these activities, dumpster diving has a certain innate appeal. "Living off the land" is satisfying to the body and soul. Dumpster diving, however, is more intense, more out-there-on-the-edge. Dumpster diving is like a

series of quick thrills. It's not mellow, like picking mushrooms. Dumpster diving is full of "cheap thrills." It is , frankly, addictive.

When I talk about "coming up empty-handed" I'm not talking about the years I lived on the ol' Hoffman homestead. As I said before, finding stuff useful on a farm is the simplest thing in the world.

Let me describe a "good" day dumpster diving. That is, a day which is "less than great." Here's the likely haul, conservatively:

*   Some food in the form of slightly bruised fruits and vegetables, or "expired" dairy products, frozen foods, bread and baked goods. It's about enough for one meal serving six people.

*   Some aluminum cans and/or aluminum scrap. It's enough to pay for, say, the gasoline we burned up detouring for dumpsters.

*   Some firewood, requiring minimal smashing and/or chopping. It's enough to provide, say, one day of heat.

*   Several magazines, newspapers, and/or a few books. You'll find more than that, but I'm only talking about the stuff you grabbed because you wanted to read it.

*   A few pieces of construction material. This may be a couple sheets of plywood, a pile of nice planks, maybe a few cinder blocks. The "firewood," with more effort, could be used for construction. Likewise, I've periodically been guilty of using quality boards for firewood.

*   A "trade" item or "use" item. It's hard not to find one great freebie, but it's difficult to predict what that item might be. It could be anything, ranging from an expensive article of clothing that's just your size to a mysteriously discarded brass candlestick to a healthy potted palm. You name it, somebody's throwing it out.

○ Animal feed and composting material. Tons of this stuff is available for the taking. Eager to head home, you grab only a two day supply for twenty-five chickens, and a few bags of leaves for winter insulation around the foundation of the house.

Now keep in mind this is the "take" on a "less than great" day. And keep in mind you only worked one hour — on your own terms — to acquire that stuff. The food alone would be tough to purchase with an hour's wages — unless you have a job that pays pretty good. But, in addition to the food, you've acquired some aluminum, firewood, reading material, construction material, a "freebie" object for use or trade, some animal feed and composting material. (Here, the "composting material" does double duty as insulation.) It's not a fortune, of course, but try buying all that stuff with an hour's wages — after Uncle Sam and the state skim their share right off the top.

As I said, we're talking about a "less than great" day of diving. I'm not talking about those FANTASTIC days when you find a month's supply of produce in one place, boxes and boxes of paperback novels, a whole truckload of lumber, or that one item you really needed and were about to purchase — like a file cabinet or a dinette set. Those are the days you yell "whoopeeeee!" all the way home, eating big mouthfuls of cheddar cheese "ala dumpster" and washing it down with pleasantly chilled fruit juice.

That's why dumpster diving — like fishing — is so addictive. The "big haul" is always just around the corner. And, managed carefully, the big haul can see you through the "dry spells" when you don't find much. But with enough skill and good territory, there's no need for you to have "dry spells" at all. And, of course, you can take the money you're saving and fix your kid's teeth, purchase a few more boxes of ammunition, or skip a day waiting on tables to try out for that part in a movie.

No, you didn't win the Florida state lottery. But compound this little advantage over the course of a week, a month, a year. It starts to add up. Pretty soon there's a confident spring in your step, a crafty gleam in your eye. People will start to notice you're doing really well for yourself.

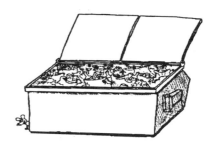

Chapter 3
Just Gimme Some Space!

Some people are poised to benefit more than others. The three big "advantage factors" are as follows:

1. Proximity to dumpsters.

2. Space (for example, a house versus an apartment).

3. Land.

Proximity

The closer you are to the dumpsters, the better. You'll waste less effort and less gasoline if the dumpsters are right around the corner. Of course, it's not likely that you'll have a little farm right in the middle of the city. A little less proximity for a little more land and space is a good deal. Our family farm was located about nine miles from a small city, and we had business in the city almost daily thanks to extracurricular activities. Three to five miles would have been better.

But proximity also means proximity to the "right" kind of dumpsters. Not all cities are created equal. The two extremes are the worst: very

small cities and very large cities. The town where I learned much of my diving *finesse* had a population of approximately 30,000. Some would consider that woefully small, but it is large enough for a maximum diving lifestyle. Keep in mind the "official" population of a city may not include heavily urbanized areas outside arbitrary "city limits." As long as the city or town in question has a good sprinkling of business and residential areas, it has potential. I've made some great finds in areas that consisted of "two gas stations and a bait store," but obviously you want as many dumpsters as possible.

The other extreme, however, is also very bad. Very large urban areas attract vagrants and, frankly, competition. I've seen grocery store dumpsters in big cities where vagrants literally sit around and wait for stuff to be discarded. There is still plenty of opportunity in a big city, but you can count on competition where food and readily salable materials are concerned. Some homeless individuals "migrate" south during the winter, or check themselves into mental hospitals, so you will note more or less competition depending on the season. Fortunately, suburban areas are just like small cities. Homeless individuals tend to stick close to the soup kitchens, blood plasma centers, and downtown missions in the inner city.

The homeless aren't the only hungry hordes that may provide competition. Towns devastated by the loss of key industries will naturally have more competition and more "interception" of dumpster-bound wealth by, for example, sharp-eyed store employees. (You'll learn how to *be one* later.) In general, any place with plenty of needy individuals spells C-O-M-P-E-T-I-T-I-O-N.

Homeless "competition" is often no competition at all.

The U.S.-Mexico border, for example, is probably the most heavily dumpster dived area in North America. Every day legions of Mexicans cross the border — legally and otherwise — and raid every available dumpster. Even dumpsters in secluded, upperclass neighborhoods are repeatedly rummaged. Dumpsters several miles from the border are fair game as the divers range far and wide for *gringo* garbage goodies. The dumps themselves are scavenged daily, as well. The fruit of dumpster diving labor is available for reasonable prices on both sides of the border, in numerous flea *mercados.*

And, with the domestic oil market in a shambles, the rest of the economy uncertain, the border states full of lean and hungry Hispanics, well, you can count on competition even hundreds of miles north of the U.S.-Mexico border.

Also count on some degree of competition in all the poorly-managed "goober belt" states. The so-called "rust-belt" is an example of an area devastated by the loss of key industries, also. Count on competition in proportion to the economic desperation in your particular area.

However, even in areas with lots of competition, dumpster diving can be profitable. I've made some of the my best finds in one of the largest cities on the U.S.-Mexico border, where 25% of the population lives below the government's arbitrary "line of poverty." Italian made leather jackets, paperback books, even that bag of coins were dived *in the very teeth of fierce competition.*

Any city with a sizable population of poor, desperate people is going to have a lot of competition. For example, there's one small city in northern Minnesota that *should* be a dumpster diver's paradise. It has a campus, an assortment of industries, and plenty of middle class residences. However, it's right next to an Indian reservation. Enough said.

Diver friends in Alaska and Florida tell me competition is fierce from native Alaskans, and Cuban and Haitian refugees.

So, in general, a mid-sized city or suburban area is best. It's big enough to have an excellent selection of dumpsters but small enough to avoid a sizable population of homeless individuals. Be aware of the economic trends in your area.

Plenty Of Opportunity

I'm addressing the issue of "competition" here and now simply because I wouldn't want you to become discouraged if you reside in a competitive territory. So much potential wealth is being discarded on a daily basis that it would take an army to cart it all off. Even in a competitive area, diving can be a profitable enterprise with only a little extra effort.

There's a reason for this *besides* the fact so much stuff is discarded. I call it my "Dumpster Diving Is Like Crime Theory." It goes like this: Crime doesn't pay because most criminals are not very bright — merely desperate. If an intelligent, careful person consciously breaks the law the odds are he or she will succeed. Dumpster diving is the same way.

In short, "competition" from the homeless is no competition at all. First of all, many homeless people rely primarily on soup kitchens, panhandling revenue, and various freebie "outreach" programs. What the homeless seek in a dumpster, when they *do* scavenge in a dumpster, is less than what *you* seek.

Most homeless have no capacity for the long-term storage of perishable goods. After all, there are no freezers under bridges and sustained cold weather drives the homeless southward or *inside*. So if they find some frozen t.v. dinners, for example, they will probably take an armful and leave the rest untouched. They aren't interested in building up a surplus or making lots of structural improvements to their shelter. They don't have animals to feed except for, sometimes, a dog or cat. They can only carry off a small amount of firewood. Sometimes a homeless person will obtain a couch or chair for their own use, but they won't carry off most large items. If they find something they can quickly sell, they'll obtain cheap alcohol, crack, or glue to keep themselves up for a day or so.

In fact, most homeless people are so busy drinking and dodging their own sad delusions that you can discount any spirited competition — except where aluminum cans are concerned. Aluminum cans are easy to convert into cash for the purchase of food and abusable substances. I've lived and scavenged with homeless individuals while writing news articles, and I watched in amazement while they ignored household goods, books and blank checks, and interesting documents in favor of aluminum cans. Can collection is also favored by many of the "marginally employed."

Yes, by the way, it's sad that some people are homeless. Lots of things are sad. It's sad our so-called "elected" government runs this country like a maxed-out "Discover" card and has regulated our once-vibrant economy into the damned grave. *That's* sad. Look out for yourself and your *kids*, not people who live in cardboard boxes because they can't cope with the cold, hard universe.

A volunteer at a homeless shelter once told me that some of their most loyal supporters — people who went out and purchased boxes of groceries and such — sometimes end up homeless themselves. Something to consider. Today's wealth is quickly squandered and tomorrow brings unexpected changes.

The homeless are not a serious source of competition except where they are numerous. There is plenty of room for peaceful co-existence. Sometimes you can even ferret out "hot" dumpsters by noting the habits of one or two far-ranging homeless individuals. Periodically I'll arrive at a dumpster and a homeless person is already there. Under certain circumstances I will immediately move in to scavenge, too. If the other person does not timidly leave, I'll simply smile and proceed to rummage. When I find something I hand over a "cut" to keep the peace. I determine the cut. Some homeless individuals with bipolar mental illness are chronic packrats, and will grab far more stuff than they can reasonably use or even more than they can carry. Hence the saving of "useless" plastic bottles and cans by "crazy old bag ladies."

Some of the homeless are too feeble to climb in the trashbin, and my assistance is welcomed — sort of. You may not understand the "word salad" of homeless speech. Don't worry about it. Smile and "talk friendly." Don't attempt this, however, unless you are reasonably confident of your ability to defend yourself.

I'll explain more about "diving etiquette" in Chapter Six.

In some ways the homeless are a dumpster diver's best friend. As long as the stigma of "desperate poverty" is attached to dumpster diving, you won't have to share the goodies with everyone else.

Here's the competition you should worry about: people like yourself. People with big dreams and an attitude of grim determination. It's ironic that this book could put equally deserving people in competition with each other. Well, such is life.

At present, America's dumpsters are like a mountain of valuable minerals with only a few wild-eyed prospectors picking at the edges. Best of all, *not everyone is looking for the same things.* There's plenty of room for more prospectors... for now.

A True Life Example

Some months ago, I noted that the dumpsters around my apartment complex were being repeatedly raided by individuals driving trucks registered in the *Chihuahua Frontera* district of Mexico. These were family groups. For example, I noted a husband and wife, mother and two kids, father and son. Their dress was not fancy but certainly clean and adequate, their trucks the same. They were after aluminum, household items, clothing and *primo* quality building materials. They were selective and intelligent about discarded food and other consumable items. They grabbed books but turned their nose up at most other printed matter. They spoke Spanish and dived confidently, without embarrassment. All the same, they moved with efficient haste and wouldn't raid a dumpster while apartment residents were standing in close proximity.

At different times, I approached each of these dumpster divers with a friendly smile and offered them some second-rate items I was planning to donate to the Goodwill. This was solely for the purpose of obtaining information, NOT to encourage them. While exchanging small talk (I speak limited Spanish, they spoke limited English), I took a good look at the back of those vehicles. These people clearly knew what they were doing. I figured I could learn a trick or two from these divers.

Observing "the enemy" at close range, I also learned their weaknesses. They preferred to dive in the late morning or early afternoon. I suspected they had children returning home from school and, like most families, preferred to be home in the evening. Also, the international bridges are dangerous after sunset.

As a rule of thumb, *most trash is discarded in the evening.* People take out the day's household trash after supper. People start a cleaning project around noon and toss out a pile of stuff in the evening. Employees let discards pile up while attending business and toss the stuff out at the conclusion of their nine to five shift. Like most rules of thumb, this one has limitations and numerous exceptions, but you will do well to remember it and act accordingly. By diving a few hours after sunset, I managed to secure "first pickings" in a highly competitive dumpster diving zone. In a few months I noted the competition had "thinned out" by about fifty percent.

We'll look at this issue a bit more in Chapter Six. But, remember, if you are smart and motivated, competition will not be a problem. CONCENTRATE ON WHAT *YOU* ARE DOING and don't play unnecessary paranoid games. Unless, of course, you happen to find something like "The Pentagon Papers." Then you should be paranoid.

Space — The Near Future

Exciting things are happening every day in space. Space is the future.

No, not outer space — *human* space. Every day capitalists are making a killing selling *space.* Rental storage areas. Little bitty apartments. Parking spaces. Bus station lockers. Clever concepts for shelving, stuffing, stacking and storing human belongings. But sadly, *space* is often a squandered resource.

Friend, I sometimes try to imagine where this nation's economy would be if we used our resources wisely instead of foolishly discarding our wealth. Not only does America throw away *megatons* of potential wealth, but the parasite government lets millions of acres of land remain idle. I'm not talking about our beautiful park system (better run by private industry), but *tracts of federal land people should be using for profit.* The government also allows tons of subsidized agricultural products to rot in warehouses and *gives away* your bloodstained, sweat-soaked dollars to backward, ingrate nations who starve and torture their own citizens. These things bother me. But another thing Americans waste is *space.*

Personally, I like to fling my pants at the foot of the bed where I can find 'em in the morning. I despise people who alphabetize their clothes by designer label or can't let a few dirty dishes sit for a day. One the other hand, I despise people who let the messiness of daily living pile up waist-deep, with no attempt to organize.

Both extremes are bad. But, having closely observed individuals on both ends of the spectrum, I draw this conclusion: an organized creature is more likely to survive and thrive than a disorganized creature. You're not going to accomplish a damned thing if you waste several hours a day just looking for your stuff.

Confronted with their own messiness, most people assert they don't have enough *time* or enough *space.* Unfortunately, the less time and space you have *the greater your need for organization.* Americans waste their time and their space in the same extravagant manner they waste their *stuff.*

I won't address time here. I will quickly address *space,* the dumpster diver's best friend... or worst enemy.

Case In Point

A few years ago, my wife and I went apartment hunting after I secured a job in a new city.

We found a reasonably priced apartment complex in a really nice area — only three blocks from a minimall with excellent diving potential.

The apartment manager explained she had one unit in our price range. The residents of that apartment were paying a fee to slip out of their lease and move to a larger apartment in the same complex. "They just got married," explained the manager. "They received all kinds of gifts and the apartment has become too small for their needs."

Rather than examining a "simulated apartment" (yeah, right) I insisted on viewing the actual unit, warts and all. Eager to make the sale, the manager called the residents at the "too small" apartment and arranged a tour for us. Possibly embarrassed by their mess, the couple occupying the apartment dashed out for a burger while the manager ushered Tina and me inside.

Yes, the apartment looked too small. *Way* too small. A couch capable of seating a rugby team battled for space with chairs, a coffee table, a home entertainment center. Albums and videotapes were stacked and shelved and scattered everywhere. Plants thrust themselves out of the cracks between the battling pieces of furniture. A half dozen lamps and track lights illuminated the scene of conflict like flares. A distractingly large bowl of fake wax fruit primped amid the chaos like a M*A*S*H* unit "Klinger." Somewhere amid the bloody mess were knick-knacks, baskets, vases, "coffee table books," a t.v. remote, a cordless phone, and oodles of fragile-looking female junk I could not readily identify.

The furniture was too big for the room, and would have been better suited to a family with kids. The couch, in particular, was *too damned big.* But plenty of other space was wasted, as well. Plants could have been attached to the ceiling in hanging planters saving floorspace. The videotapes and albums could have been better organized, too. The space at floor and waist level was a virtual traffic jam, but more efficient shelving could have solved the problem. There were so many "decorative" items that the actual stuff of daily living was pushed aside. Delicate glass thingybobs staked out vast tracts of choice shelf space while attractive leather-bound books and nicely framed photos formed squatters colonies on countertops. It was a shame. Worse yet, when people like this obtain more space they clutter *that*, too.

The kitchen, if anything, was worse. Every device for perking, popping, straining, mixing, blending and microwaving was jammed into the kitchen haphazardously. I saw my wife's breath stop as she noticed the biggest, prettiest, most expensive spice rack she had ever seen. None of the spices had been *used*, of course, and dirty dishes in the open dishwasher were not accompanied by pots and pans. *Decorative* copper pots and pans hung on the walls in a high state of shine. All these expensive kitchen gadgets served the same purpose as those knick-knacks.

"What are you *doing*?" asked the manager, as I pulled out a small, powerful flashlight and proceeded to look under the sink.

"Checking for water damage, rodents, termites, things like that," I replied, with a friendly smile.

She gave me a tight smile and a look of grudging respect. How could she, an apartment manager, resent a tenant who was careful and conscious of property?

I proceeded to inspect all the cupboards and corners in the kitchen. I ran the faucet and the garbage disposal, inspected the fridge and pushed a few buttons on the dishwasher.

"A dishwasher!" my wife said, giving me a secret look.

I knew what that look meant. While the manager babbled about the wondrous efficiency of the dishwasher, I give my wife a wink. Later we would ask to have the dishwasher removed (obtaining a receipt) and would establish a little pantry in the space, instead. A dishwasher does the following: 1) wastes energy, 2) wastes water, 3) wastes detergent, 4) never gets dishes totally

clean, anyway, 5) eliminates a task which is, in my opinion, kind of relaxing, 6) requires much more space than it's worth.

Think about it. Two feet by two feet by three feet is *twelve cubic feet* of valuable kitchen space. Some models are even larger. And for what? That water-wasting, soap-swallowing, space-squandering Maytag monstrosity doesn't save you more than twenty minutes a day — and you lose *that* in water, electricity, soap and repairs.

Even as I examined every foot of that apartment for potentially unpleasant post-lease surprises, I couldn't help but notice the poor use of space by the current residents. Mentally, I was rearranging things — putting up a space-saving rack on the inside of a cupboard door, for example, and removing foodstuffs from their bulky packaging.

The small bedroom of that apartment was worse than the living room and kitchen combined. Boxes of personal records spilled out of closets crammed with out-of-season clothing. Sports equipment, hunting gear and an exercise bike occupied precious living space. (An exercise bike, by the way, makes a dishwasher look like survival gear.) I peered under the bed and found empty space — except for a few sex toys. Bookshelves, adorned with more prissy decorative objects, lined the walls with thousands of volumes. Why, I wondered, didn't the young couple just sell some of those books?

"Well, that's about it, I suppose," the manager smiled.

"Not quite, " I replied.

Standing on a chair, I removed a ceiling panel and inspected the crawl spaces above the kitchen, living room and bedroom. And my eyes just about popped out of my head. No, I didn't find a nest of rats. I found *another three rooms*.

Up to this point, I had also considered the apartment too small. But, I figured, we would organize our possessions and survive, until we could afford a better place. But these crawl spaces were *three feet high*. Subtracting the space occupied by pipes and heating ducts, there were still hundreds of square feet of neglected space.

My mind exploded with the possibilities. Shining my flashlight around, I imagined an "herb" garden, complete with hydroponics and ultraviolet lights. I pictured a hiding place for urban freedom fighters, a production center for radical anti-establishment propaganda, a smuggling operation, and various other covert activities.

As it was, we used the extensive crawl space the way the previous occupants *should* have used them. We stashed personal records there, always careful to keep the area free of paper-chewing vermin. We used it as a storage area for out-of-season clothing, firearms, recreation equipment, and "sensitive" items we wouldn't want to leave, say, under a bed. But most of all we used it for the storage of bulky dumpster dived articles until we had enough stuff for a trade or sale to a consignment store. We managed to keep more stuff in that apartment than the other couple, without being crowded.

Finding Your Own Space

Why all this talk about *space?* You *must* be organized and efficient, or a deluge of dumpster dived goodies will be more of a bane than a blessing. You *must* be intelligent and practice self-control, or you'll simply be a much *poorer* and more *tired* pack rat.

Ask yourself if dumpster diving is really for you *right now* — or could you benefit more by organizing your current resources, budgeting your money more wisely, and controlling your habits? I've met fairly efficient divers who spend their dumpster derived wealth on alcohol and compulsive gambling. Their kids will grow up to curse them. One day their kids will be sitting in some state hospital, attending group therapy, and they'll say, dramatically, "My father made us *eat out of garbage cans.* Boo hoo hoo."

If you can't control *yourself,* how can you hope to have control of the world around you? What good will it do to increase your wealth by 25% if you waste it and need more? If your kids hate you? If your wife leaves you? Consider the lotus which grows from the decaying muck of a swamp, deriving its beauty from that muck. The lotus is an example of *perfect design.* Be like the lotus, Trashhopper. First dive within yourself.

Do yourself a favor. Read a book on time management instead of watching the Disney Channel. Time is money. And consider these ideas for better use of space. Space is money, too.

- *Crawl spaces, attics, and basements.* Working just an hour a day, listening to the radio instead of watching t.v., you can convert these spaces into efficient storage areas.

- *Garage.* Park the car in the driveway with a tarp over it and your favorite anti-theft device. Make that garage a workshop or den.

- *Yard.* Build a subterranean cinder block storage facility. You'll still have a whole yard, only there will be a small, hidden building there, too. When you think about *digging,* you'll see you have more space than you ever imagined. (Remember, however, that underground utility pipes and cables can kill you.)

- *Under the bed, beneath the couch and chairs, and in the space where you used to have a dishwasher.* Consider hooking things to the ceiling. Install racks on the back of cupboard doors and other neglected spaces. Obtain a used bunkbed for each of your kids and convert the whole "lowerbunk" area into storage space. College kids call this a "loft." It works for mature individuals, also. As times get tougher I predict we'll see lofts become more popular. Beat the rush and start a trend in your area.

- *Other people's spaces.* Consider all the places you might "borrow." Once, I had a complete garden on somebody else's property. And I once knew an individual who lived in a terribly small apartment but did a lively trade in various types of discarded property. He simply used space at a building owned by his employer. Brain storm a bit about the space around you and how to access it. You may find yourself tunneling into your city's concrete guts.

- *Car trunk.* You *should* drive a truck with a removable "topper," but we'll discuss "equipment" later. All day long we haul around empty space in the trunks of our cars. Load up that space with lightweight items you wish to sell. Does this cost you gasoline? Hey, you don't calculate the "extra gas" when you have a 180 lb. passenger. Riding your brake and revving your engine is how you burn up gas.

- *Rooftops.* Weather is a factor here, as well as pollution. But consider the possibilities for gardening, your own instant "deck," etc. Pry those windows open and walk in the sunshine. Don't fall through a decayed roof, however. As times become tougher, cities will appear greener and greener at the skyline.

Hopefully, you've started thinking about your own space. The more space you have, the better, particularly where messy items like car batteries are concerned, or bulky items like aluminum siding. If you live on a farm or something like a farm, you're in luck. You can salvage and store lumber until you have enough to sell for some quick cash. You can drag home a big piece of plate glass and store it somewhere for years until you need it. The more space you have, the more you can profit by dumpster diving.

And, remember, when it comes to dumpster goodies: convert, convert, convert. Don't leave the stuff sitting around like Federal Reserve Notes stuffed into a mattress. Put your wealth to work for you. If it's not part of your personal stash of items for future use, convert it. Even using the cash to buy soda pop and pay the cable bill is better than letting boxes of dumpster loot sit around, crowding you out of house and home.

I once met a women who saved all her magazines and her children's old toys, among other things. She showed me some antique guides, once, and pointed with pride how much her old toys and old *Life* magazines were worth.

However, she crowded her own children out of house and home with her crap collecting. She even demanded space in their little rooms for long term storage of moldy old clothes. Years later, the mere mention of her habits can set those children off in explosions of scarlet-faced rage. They are defiantly wasteful of material possessions, tossing out a shirt and matching pants, for example, rather than fixing one button. (I *love* diving their trash.) And the old woman died damn near penniless from an infected tooth, still surrounded by thousands of dollars worth of rusting, moth-eaten "antiques." She was still waiting, I'm told, for "hard times."

Land, Ho!

Once, everyone was directly dependent on what he or she could hunt, grow or gather from the land. After a while some people became more dependent on their own ability to produce — let us say — arrowheads. These people became "once removed" from direct dependency on the earth. Nowadays, the majority of Americans are six or seven "times removed" from direct dependence on the earth for their own livelihood. Blizzards, droughts, and crop failures come and go with little noticeable impact on our lives. And that's good.

However, sometimes we are so far removed from the land that we become abstract idiots. Americans have become like domestic turkeys, who allegedly look up in the sky and drown during a rainstorm. But, a funny thing has happened. Even as our civilization becomes more advanced than any preceding civilization, people long for contact with the land. People wish for "a little farm," and "old fashioned values," and "self-sufficiency."

What's weird about this primal urge to return to the soil is that people approach it the way they approach most city problems: they attempt to "purchase" self-sufficiency. And once they look at the retail price of garden seed, tools, livestock, fertilizer, building materials, and so forth, they turn pale and drop the whole idea.

This doesn't need to be the case. Dumpster diving can be your cheap ticket to self-reliance. Everything your self-sufficient, organic little heart desires can be found, eventually, in dumpsters. And plenty of the stuff in those dumpsters can be converted to cash or bartered for the things you need. Take 50% of that money you were going to spend on garden seed, building supplies and tools and *put it in your favorite investment.*

So, next to *proximity* and *space, land* is another factor in the maximum diving lifestyle. Take food refuse, for example. On the ol' Hoffman homestead, we kept a substantial number of pigs, chickens, goats and rabbits happy and well fed with discarded food. You should see how pigs go hog wild for cold pizza. The money we saved on animal feed was immediately invested in stuff like dental work.

Some purists would argue this is not "really" self-sufficiency, since you are "dependent" upon things discarded by city dwellers. Don't worry about it. When the Eventual Economic Collapse (EEC!) hits, you'll have a huge supply of materials socked away, and the "purists" will be weaving their damned birchbark baskets.

In the not-too-distant future, I believe small scale farming will take place in the very heart of the city. Encouraging developments are also taking place with hydroponics and other "futuristic" methods. Perhaps the day will come when all those crawl spaces are filled with ultra-advanced home hydroponics modules. In the meantime, however, the only plants worth hydroponics and ultraviolet lights aren't exactly legal for human consumption.

Finding your own *land* to plant is similar to finding your own *space* to use, only planting onions and potatoes here and there is considerably less obtrusive than storing piles of lumber on somebody's vacant lot. A "compost pile" or "compost box" may be rather obvious, but it all depends on your own situation. And, done right, it doesn't need to be time consuming. The point of "guerrilla gardening" is to produce *food*, covertly if necessary, not to create a symmetrical, weedless little plot for display purposes.

Of course, this is an activity which carries no absolute guarantee. Once, I had a delightful garden started near a house I was renting for the summer. Everything was going perfectly, and I figured a bumper crop of veggies would reward my efforts. Then — rabbits! All I managed to produce that summer were pears and a few flowers. Fortunately, I ate enough rabbits to justify the "bait" I had planted.

Knowing that I can produce my own food in the midst of unforeseen circumstances gives me a priceless sense of security and self-worth. In peacetime, armies practice for war. Likewise, you should "practice" for unforeseen circumstances. When you *need* a garden it's too late to plant one — or learn how.

Once, I decided to try a "self-sufficiency experiment." For one month, I ate only those foods I could hunt, trap, grow or gather. I continued to dumpster dive and to "intercept" food at my place of employment, but I didn't consume it. I simply "stockpiled" the salvaged food. I even developed an elaborate system of labeling, making sure I consumed only "experiment" food.

And what did I learn from my experiment?

1. A garden can save your ass.

2. You may *think* you know wild edibles when you see them, but there are more than you imagine. Learn about wild edibles before *hunger* becomes your motivation.

3. If you see some animal you can eat, *kill* it and throw it in the freezer. Learn how to skin a squirrel *now*, before you have three

hungry, wide-eyed kids watching your confused efforts.

Yes, dumpster diving is a great food source. However, a dumpster based diet is full of processed foods and pesticide treated produce. Even in the middle of the city, you should explore the possibilities of trapping fresh meat, gathering wild edibles, and growing your own produce. Dumpster diving can help you obtain the materials or cash to engage in these efforts.

For example, some people like to drive to the country and pay to "PICK YER OWN APPLES." Some people even pay for a garden plot in the country. I know, I've rented these plots. The auto mileage and effort sometimes costs more than the value of the produce. However, these activities look downright profitable compared to, let us say, a local amusement park. It's more memorable, too.

Remember, kids raised on a steady diet of television, amusement parks, Nintendo and processed food will not change readily. *Be patient.* You raised 'em. The apple doesn't fall very far from the tree, so to speak. Sometimes the child who cries and screams to visit "Boffo Funland" will turn into a kid who begs to go work in the garden. Let them plant things, pick things and "shoot" the water hose. This is much more satisfying than watching Ninja videos with them. Point out pretty growing things, big things, unusual things, don't critique their every little move. And wean 'em off television. It fries their brains.

Plant something. It's a life affirming act of defiance, and it will do you good. If you learn something, if you eat the fruit of your labor, so much the better. And all the fertilizer and composting material you need is in dumpsters.

Consider these "alternative" farms:

- *Backyard.* Consider raising a few chickens or rabbits. If you are itching to fight city hall, this is a great way to provoke the bureaucratic bastards. But, if you are keeping a low profile, consider a garden. Even the narrow strip of dirt around the house is suitable for cultivation. Improve the soil with compost and consider sunlight, moisture and temperature conditions before planting.

- *Look into root vegetables.* Sure, it would be great to show off a big squash, but squash require lots of soil, space, water and sunlight. If a rabbit or groundhog gets near it, wave bye-bye. (Young groundhog, gopher, and woodchuck, by the way, taste like pork. Try it.) With root vegetables, you can figure out that the *Rodentia* are conducting raids before extensive damage takes place.

- *Vacant lots.* You may think somebody would come along and grab your veggies, but that isn't usually the case. Carrots, onions, and potatoes look just like weeds to most people. Improve the soil and use foundation remnants for composting. Look for city water hook-ups to water your crop. Check carefully to insure hazardous chemicals haven't been dumped on the lot.

- *Right-of-ways, grassy medians, ditches, underutilized public land, and so forth.* Obscure laws exist in some areas which allow you to plant gardens or harvest hay and certain wild edibles openly and legally on some "public lands." Deadwood, or even standing trees can be gathered, cut or transplanted legally if you ferret out the correct procedure. (Don't take the word of the very first paper-pusher you encounter — in anything.) But why bother? Just go ahead and do it.

Be careful that your compost can't be mistaken for garbage and somebody busts you for dumping without a permit or some other hocus-pocus law. Plant some flowers and fruit trees. This will make you look real good if you encounter a legal hassle or wish to provoke one.

- *Somebody else's land.* Simply look for "private" land which isn't closely watched and can be used without hassle. Empty homes with large yards, for example, or vacant commercial sites. You don't have to reside near the garden, either. I once knew a meter reader who planted veggies at homes which were vacant. Some of the homes were occupied by the time his veggies came to maturity, but he always had a fairly large crop in the fall.

- *Somebody else's farm can be your farm.* Once, my mother was planting "windfall" squash seeds on semi-cultivated areas around our farm buildings. These are seeds we saved from rotten squash we found in dumpsters. A huge gust of wind scattered some of the seeds in the neighbor's field, where he had just planted soybeans. We later discovered several hundred pounds of squash amid the soybeans. We figured the squash plants were not consumed by the rabbits because we hunted the rabbits constantly. After that, we made a point of planting squash amid our neighbor's crops, grateful for his tilling and irrigation.

Plant something. It's anti-establishment.

What's In It For Me?

We have been examining the relative advantages of *proximity*, *space*, and *land*. You know your own situation best, and you can adapt accordingly. But, in addition to the three previous factors, an all-important "fourth factor" can change the whole picture: YOU.

Are you intelligent? Are you motivated to work for your own interest? Would a big box of food make you happy to the point of delirium? You may have a few diving advantages, or every advantage, but whether or not you benefit depends completely on YOU.

Some people are poised to benefit more than others, and I have previously described these people as the "lean and hungry." Whatever the world hurls at them they manage to adapt, overcome, survive and, ultimately, thrive. And when the lean and hungry have something valuable thrown their way, by God, they do something clever with it. Introduced to the tre-

mendous windfall of dumpster wealth, the lean and hungry find ways to use unexpected dumpster dived *objets d'art*.

Somebody who is not one of the lean and hungry might look in a dumpster and say, "Duh... let's see here. Cans of fluorescent spray paint. Nylon rope. Crash helmet. A box of road flares. And a female department store mannequin? Aw, shucks. I was hoping for porno mags. Now what can I do with this crap?"

The lean and hungry individual, however, feels the hair stand up on the back of his neck. No, he isn't sure what he can do with the stuff, but he knows immediately that he'll think of something. The lean and hungry individual senses the tremendous potential, if not the exact application and value. He or she will grab what life issues him or her and use the hell out of it. By God, the lean and hungry will adapt and overcome.

Oh, What A Feeling!

Before I set out to dive dumpsters, a tremendous feeling of excited anticipation makes me tingle from head to toe. You never really know what you'll find. I've recovered a box of Soviet medals, military skills manuals, a fossil collection, war mementos, hundreds of old *National Geographic* magazines, walrus tusks, autographed photos of movie stars, two-headed baby pigs in jars of formaldehyde, sets of encyclopedias, and so forth. Needless to say, I've found several kitchen sinks. Another exciting thing I frequently find is information, to which I've devoted Chapter Twelve. Take, for example, a list of fire safety violations at my least favorite local business. An anonymous letter here, a photocopy there, and next thing you know somebody's whole week is ruined.

Was I looking for this stuff? Not specifically. Did I make use of the stuff when I found it? You better believe it. I used it, sold it, bartered it, and/or stockpiled it as I pleased. And you can do the same.

Your ability and, more importantly, your LEVEL OF MOTIVATION to adapt and overcome is the all important "fourth factor." Ultimately, it all depends on YOUR brains, YOUR will, and YOUR initiative.

Military gear discarded in an apartment complex, including live ammo and bullet-pierced Iraqi license plate. All this was picked up in one dive.

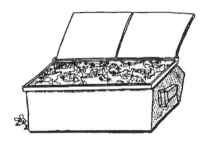

Chapter 4
Is the Wealth Really There? (Yes!)

Material possessions are easy to take for granted until you don't have something you need and can't acquire it. I grew up elbow to elbow with people who experienced stark poverty. I saw little kids develop pneumonia for lack of a warm winter jacket. Even for the well-to-do, property represents safety, health, well-being... LIFE. For example, I could hock my television to buy my wife medicine. So even "non-essential" property contributes to your overall well-being.

But many Americans don't have a "safety net" of property... they have debts. Amazingly, they discard their goods and tenaciously accumulate debts.

Dumpster diving clears your head of a lot of bullshit. Just as I rarely go diving without finding something, I rarely return without learning something. Mostly, I learn people are dissatisfied, confused, and waste a great deal of energy. The average American's head — I'm convinced — is a lot like his dumpster. There is *some* good stuff in there. But there's a lot of crap, too. Your average American can distinguish between good stuff and crap — but just barely.

He gets all confused sometimes and discards some of his good stuff, believing it to be crap.

That's where YOU, the extraordinarily clear-headed dumpster diver, enter the picture. Everything you salvage — whether for personal use or sale — adds to your stockpile of wealth, compounds little advantages to your personal economy.

Is the wealth really out there? Yes, yes, YES! Americans represent only six percent of the world's population, yet we produce *almost half* the world's waste. And I've chosen the word "waste" deliberately. It's wasted Wealth.

I've searched for statistics on American garbage and found most to be confused, cloudy, and contradictory. The government is more concerned with counting other things. But here's my best summary of TOTAL WASTE.

- Ten to fifteen pounds a day for every man, woman and child. That's almost two tons a year. So if two dumpsters sit outside an apartment complex containing three hundred and sixty five, every day each of those dumpsters will "represent" the yearly waste of your average apartment resident. Now, do you suppose that over the course of a year the average person throws out one or two things you could use? Well, every day one or two useful items are sitting out there is those dumpsters.

- Half the waste is packaging, mostly plastic and paper. So, by basically ignoring this crap, half your work is done right there.

- Of the remaining fifty percent, thirty percent is "organic" in nature, i.e. food waste, Christmas trees, lawn trimmings, dead parakeets, etc. Good to eat, good to feed animals, or good for fertilizer. But we choose to bury and burn the stuff like it's mutant ooze. METAL makes up roughly ten percent, and another ten percent could be considered MISCELLANEOUS. For example, framed pictures, still-working Christmas lights, boxes of telephone equipment.

Of course, that is a very broad overview of a complex and beautiful mountain of stuff. A grocery store dumpster, for example, is about twenty percent organic and about eighty percent plastic and paper. A bookstore is almost ninety percent paper, in the form of discarded books and cardboard boxes. A restaurant is a mix of "paper" and "organic," in widely-varying ratios, while an appliance repair shop could be almost fifty percent miscellaneous. Still, there are plenty of times when the grocery store dumpster is full of discarded sheetrock, or the bookstore dumpster contains the discarded remains of an employee potluck dinner.

Quality varies considerably. I wouldn't consider personally consuming ninety percent of residential food waste. But I would consume almost fifty percent of grocery store waste. Looking at the broad statistics establishes that a lot of stuff is out there, but provides little detail.

For detail, ask a dumpster diver. I'm telling you the discarded wealth is enough to make you *weep*. I have brought people along on my diving forays and made them converts in a single night.

The question which always comes out of their mouths is, "How could somebody just throw this stuff away?"

Let's explode a stupid myth. Here is the myth: People throw things away because the stuff isn't good anymore.

WRONG. People throw things out for a variety of convenient reasons. They are moving and can't pack all the "junk" they saved in closets and attics. Somebody died and left a house full of "crap" to dispose of quickly. Somebody gave them a gift they didn't want. Somebody broke their heart so they threw away a whole box of "bad memories." People throw away clothes which are "out of style" or which they have "outgrown" or which have one little spot or tear. People toss out items which are nicked, scratched, bruised, dented, or otherwise "damaged." The reasons people discard things are as complex as the reasons people purchase things.

Here's another stupid myth: If the stuff had any value it wouldn't be discarded.

Yeah, tell me that while I jam fifteen dollars "cash money" in my pocket. I've seen people throw out boxes of clothes that have been washed, pressed, and neatly folded. I've acquired music tape collections in perfect condition. People throw things away mostly because they happen to be idiots.

One of the biggest reasons dumpster diving is so profitable involves the American obsession with "cleanliness." You'd think we were all raised in oxygen tents. Most Americans won't purchase a "blemished" vegetable and most stores won't attempt to sell one. Americans seem to believe they'll puke their guts out and die if, God forbid, they should be forced to eat bread with a little mold on it, food from dented cans, or a slightly wilted vegetable. (Never mind the fact that we are all eating ground up insects, rat droppings and manmade poisons in all the perfect-looking food.) If I could take the food I salvage from American dumpsters and sell it in Russia, I'd be a *millionaire*. And you can bet the

Russian housewives wouldn't bitch about a bruised vegetable or a dented can.

I've been eating dumpster dived food my whole life, and I have never been seriously ill as a result. I've never had food poisoning, and I've eaten from cans that were so bloated they looked like squash. Two years ago I jumped jobs and had to pass two rigorous physical exams (including a PT test) within a space of two months. I passed both with flying colors and was declared "in perfect health." Oh, I'm sure I encounter more germs than the average person... since I work in a hospital.

Being a proficient dumpster diver provides me with a priceless sense of security and confidence. As a teenager, I would ride my bike all over the midwest, not bothering to bring food or money. I knew very well I could take care of myself.

Recently, I found a coupon for a free oil change. It appeared that somebody had been cleaning out their desk before Christmas vacation, and the coupon was discarded amid a lot of other papers. I was lucky to spot it; you develop "diver's eye" with practice. The coupon, as it turned out, would expire the next day (December 31). So I had a small "window of opportunity." I finished my "route" and drove approximately one mile to the service station.

The deal on the coupon applied to only members of a particular group, but I managed to *finesse* my way through that part with the manager. My clean cut appearance and sincere attitude helped quite a bit. Besides, most managers don't give a damn. "Freebies" are for the purpose of showing off a product or service, getting you as a regular customer and convincing you to buy "extras." The stores are usually compensated by the home office and would gladly give makeovers to winos if compensated at a slight profit. The service station manager agreed to "squeeze me in" some time in the next two hours, then bored me for about five minutes with a presentation about some "miracle" oil additive. I played along by asking a few questions, grinned right through the "hardsell," expressed regret but gave a firm "no." I placed my ignition key on the counter and told him I was *sooo* happy he could "squeeze me in."

I figured I could kill two hours dumpster diving in this unfamiliar neighborhood. It sure beats "window shopping," though I like to combine the two. Looking through a store's dumpster after browsing the store is like seeing your favorite actress without her makeup, wearing torn blue jeans and eating fast food. Dumpster diving is a brutally real way of examining the world.

The first three dumpsters I checked were interesting, amusing little bins... but basically "dry wells." However, the fourth dumpster, outside a sprawling apartment residence, was "warm." Somebody had received one of those gourmet "goody baskets." You know, the kind with fake grass and nicely arranged meats, cheeses, etc. He or she had eaten all the meats, but had ignored half a dozen individually wrapped cheeses, a small box of melba toast, and a little jar of apple jelly. I didn't have room for the basket in my dumpster dived "Adidas" bag, so I decided to just grab the food and throw the basket back. That was when I noticed half a dozen current magazines, mostly women's magazines.

"My wife will like those," I thought.

As I grabbed, one of the magazines slid out of reach behind a discarded Christmas tree.

"Damn, " I thought. "What is that? A J.C. Penney catalog?"

I pulled several pieces of plastic-wrapped candy off the tree and sucked on one of them, thoughtfully. Experience has taught me to double check my dumpster assumptions. However, I didn't want to rummage amid that dried up pine tree for a lousy catalog. And I had neglected to bring my "dive stick," a collapsible six foot antenna off a portable citizens band radio. Looking around, I spotted a broken pool cue. I grabbed the cue and probed at the catalog. Flipping it over, I saw it wasn't a catalog, but a copy of *Swank* magazine.

"Well, flog me!" I breathed, and grabbed the magazine.

As I brushed pine needles off myself, I noticed that a utility building near the dumpster was adorned by signs and stickers reading. "NO TRESPASSING," "THIS PROPERTY PROTECTED BY .38 SPECIAL," "NEVER MIND DOG, BEWARE OF OWNER," and other macho bullshit. I decided, however, not to test my luck. About this time I was getting thirsty, so I walked over to the Diamond Shamrock convenience store. I wondered if I should break down and purchase a bottle of Coke to wash down the cheese and melba toast.

"Nah," I decided. "I'll scavenge a nice clean container, scare up some ice inside the store, find a faucet and drink water, instead."

As I rummaged through the Shamrock dumpster for one of their plastic jumbo beverage cups, my eyes popped out of my head. Amid torn maps, tourist brochures, and empty candy wrappers, somebody had discarded two cans of Pepsi — still attached by a plastic ring.

If you dumpster dive for any length of time you'll understand my excitement. I find books of blank checks and porno magazines more often than I find sealed cans of cola. This was a rare treat.

"Thrill me!" I breathed, and grabbed the cans.

Perhaps the colas were warm and icky when discarded, but that cool, shady dumpster had chilled them perfectly. I retired to the shadow of a large evergreen to drink my soda, eat my cheese, and check out the skin mag. As I crossed the intersection I noticed a homeless man holding up a sign.

"Hungry," read the sign, "Please Help."

Sampling my gourmet cheeses and checking out "foxy boxing," I reflected on the past few hours. I had acquired a free oil change, including oil filter, air filter, fluid check and vacuumed interior — retail value $21.99. I had obtained

lunch and reading material. And I had done it all on my own terms, *as I damn pleased.*

I watched the man in the intersection beg passing drivers for coins. Who was he? I wondered. Did he once have a family? An apartment? Credit cards? Did he ever dumpster dive in this neighborhood?

I couldn't help but think that he should have learned how to dumpster dive before he was "hungry and homeless." Maybe if he had saved twenty bucks on an oil change, five bucks on lunch, a few bucks on a magazine — well, maybe he wouldn't be out in the intersection humiliating himself.

Dumpster diving of your own free will isn't icky or humiliating — it's *invigorating!* When you make a great find you tell yourself, "By God, I can survive and thrive ANYPLACE!"

Of course, actually sticking your head in a dumpster is a big hurdle for a lot of people. I have difficulty understanding that point of view because of the way I was raised, and because I *know* how much fun and profit there is to be had in dumpster diving. But for those of you trying to "jump off the diving board," consider the following:

- You've been around dumpsters all your life. You've opened them, closed them, and looked inside. Some people make more money than teachers working as "sanitary engineers." They don't get sick or hurt all the time. *It's no big deal.*

- If somebody offered you some food, or clothing, or an appliance, and the stuff *looked* o.k., you'd probably take the item off their hands and thank them kindly. Even if they said, "I'm ready to throw it out — but it's still good," you'd probably do the same. Well, dumpster diving is just a few more steps up the ladder.

- Hey! I'm a college educated guy who is smart enough to write a book. I have a good job, a beautiful wife and a nice apartment. I have very few debts, and several small investments. I dumpster dive constantly and it's fun, fun, fun!

Recently, I found a big box full of food sitting next to the dumpster at my apartment complex. Somebody had apparently moved and cleaned out their fridge. Usually I can salvage three or four items in a box like this, so I poked around to see what I might find. That was when I noticed the fridge wasn't the only thing they had cleaned out — the bottom of the box was full of dry goods and canned items.

"Jackpot!" I gasped. My leather gloved hands were turning over cans of vegetables, an unopened bag of flour, bags of pasta, cooking spices, powdered drink mixes...

Grab the stuff and run! I thought. When I find a box that looks like it's filled with good stuff, I don't stand there and pick through it — I split! I grabbed the whole box and dashed to my apartment. Then I went back to the dumpster to see if there were more goodies. When I find something good in a dumpster, I keep looking until I'm sure there isn't any more left. Diving luck comes in "streaks," and you have to play those streaks like mad. If I check twenty dumpsters and find a big pile of good stuff, you can bet that stuff was found in one or two "hot" dumpsters. In a residential area the "hot" dumpsters change day to day like the "hot" numbers on roulette wheels.

I spent the next half hour sorting the stuff in that box and carefully wiping some of it off. I keep a lot of torn shirts around for cleaning up my "finds." Where did I get the shirts? As if you had to ask! I put that food away in my cupboard, thrilling over each can of split pea soup and box of macaroni as if I had just created the stuff out of thin air. Of course, *finding* the stuff is the biggest thrill, but I also enjoy cleaning the stuff up, putting it away, calculating what I'm going to do with it. I enjoy this like some people enjoy putting wax on their car and making it shine. I like putting my hands all over these items and saying "You are mine, mine, all mine."

I don't become *this* excited when somebody gives me a gift. Gifts come with strings attached; when you find something in a dumpster *it's yours.* Months later I might walk down the aisle of a grocery store and say, "I *found* a can of soup like that, once." Just as a haberdasher has a "raised consciousness" about clothing, noticing things a normal person wouldn't, dumpster divers develop "diver's eyes."

When I walk through the produce section of a store, I notice right away the cracked coconuts, the bruised melons. I notice stuff nearing its expiration date in the frozen foods section. But it's more than that. *Everything ends up in the dumpsters.* Dumpster diving raises your consciousness of the world to a greater degree than a specialized profession in a small niche of the marketplace. The other day I found an odd item in a dumpster. I didn't know what it was, but it looked like something that came off a car. Well, the next day I was sitting in my car at a stoplight when I noticed *that very item* on the car next to me. This happens thousands of times a year to me. I'm not suggesting we build a belief system around "dumpster based consciousness," but I'm convinced dumpster diving teaches me more in an hour than, say, waiting on tables.

Well, I put all the food away in our cupboards... except for one can of corn, which I ate right away. I took the useless items (messy, half-used bottles of ketchup, spoiled hotdogs, etc.) and tossed that stuff right back in the same dumpster.

When my wife arrived home I showed her the additions to our pantry. She squealed, delighted. The staples like flour and sugar were an unusually good find. It happens, but not often enough so you can count on it. Of course, the more you dive the more it happens.

That evening my wife went to the grocery store — to *shop*, not to dumpster dive. When she returned she had only a small bag.

"What's that?" I asked. It looked like a *surprise.*

"Something special I bought," she said. "Since we didn't have to spend anything on groceries this week I figured we could afford a treat."

She pulled out a jar of *caviar*.

"Oh, boy!" I said.

"You like it?" she asked. "Had it before?"

"Never had it before," I answered. "Always wanted to try it."

My wife consulted an obscure portion of her food science book. She served the caviar chilled, with cream cheese, on fancy (dumpster dived) crackers. Dumpster dived celery and lemon wedges served as a garnish.

Like most of my family, I'm crazy about fish. I thought the caviar was terrific. I decided I definitely wanted to incorporate this delicacy into my life.

"You know," I told my wife. "I'm going to like being rich."

You just never know what you might find. It's like trying to guess the name of a gnome who weaves straw into gold just for you. Guess from here to kingdom come, you'll never guess "Rumpelstiltskin." Only by sneaking up in the dead of night will you know the correct answer is, "three wool sweaters and a portable pet cage." Dumpster diving is outrageously unpredictable. You have to *do it* to believe it.

If you dive consistently and well, however, there are a few things I can reasonably predict you'll *never have to purchase again*. Here are a few of those things:

Fresh fruit and vegetables... manila envelopes... clothing... clothes hangers... boxes... houseplants, including pots... Christmas decorations... cheese... videotapes and audio tapes for recording purposes... furniture... candles... knickknacks... bread and bakery goods... most toiletry items... books, magazines and newspapers... low cost jewelry... and much, much more.

What would *that* do for your budget?

Naturally, I still purchase plenty of things. If I don't find any artichokes in the dumpster and I get a powerful hankering for artichokes, I'll buy some. But the more flexible you are willing to be, the more you can adapt yourself, the more likely that your needs can be answered through dumpster diving.

It's tough to predict what you'll turn up in your area. I didn't put "candy" on that list above because it isn't commonly found. However, I once found roughly half a year's supply of assorted candy. Be flexible, and you will answer many more needs. For example, if you aren't particular what kind of shampoo you use on your hair, *great*. If you just gotta have Pert Plus, well, you're going to end up buying it and tossing back a lot of perfectly good half-used bottles of shampoo. You'll notice that I *wanted* a Coke but was happy to acquire a Pepsi at the convenience store dumpster.

Hey! I almost forgot something important! When I dumpster dived those goodies *nobody challenged me.* In fact, I haven't been verbally confronted about dumpster diving for *almost three years.* And, on that occasion, the old hag... er, elderly lady in question told Jed and me to "go dig in the neighbor's trash, not *mine!*" I don't think she would have said anything, but we were being rather loud and taking pictures, too. So we went and dug around in the neighbor's trash. And it was worth it.

When you dumpster dive, *you become invisible.* People will walk right past you as though you don't exist. No doubt they are afraid you'll ask them for a handout or they don't want to embarrass you. It's worse when you don't "look" homeless. People don't know what to make of you. Are you looking for something? Do you work for the sanitation department? Are you mentally deranged? Pity, fear and confusion keeps them away. But the same thing keeps them from calling the police. Avoid them, especially when they need to deposit something in the dumpster, and they will avoid you and try not to think about the fact you even *exist*.

My little brother, Jed "Slash" Hoffman, calls this phenomenon "ninja dumpster diving." He loves to imitate the repressed, eyes-straight-ahead look of the non-dumpster diving people walking past. When I dive with him my sides ache from laughing as he does his "ninja" thing amid the boxes or does deadly accurate imitations of old ladies.

"*Don't look*, Rose!" he says. "My God, they're in the dirt bins! What are they *doing*? Ooooh, they found something. What is it? My God, *they're sniffing it! Don't look*, Rose!"

If these people knew I work in a local hospital and sometimes eat the stuff I find with caviar smeared on it, *their heads would explode.*

So what are you scared of, comrade? Cooties? Thar's gold in them thar dumpsters.

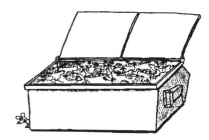

TWISTED IMAGE by Ace Backwords ©1993

Chapter 5
What the Well-Dressed Diver is Wearing

The Right Stuff — Clothing

Let's start with the basics. You want to go dumpster diving but what to wear, *what to wear?*

Well, dumpster diving is an informal event. Don't dress sloppily, but don't overdress, either. You want to project the idea that you have a home... somewhere.

Consider the weather, first. Wear layers of warm clothing or rain gear if necessary. Don't wear bright colors, weird hats, political buttons, or anything else that will make you extremely noticeable. On the other hand, don't dress completely in black with a stocking cap and gloves. If you're skulking around an alley at night *you don't want to look like a damned burglar.* Particularly, I might add, since you will probably be carrying such things as a flashlight and a knife to cut plastic bags.

This may seem obvious, but don't wear stuff that identifies you as part of a group, business, or organization. Don't wear your "Pizza Dudes" windbreaker or a sweatshirt with the word "Army." You might think this is very clever if you don't actually belong to the group whose

stuff you are wearing, but it just makes you look distinctive. It ruins that image you are projecting, makes people wonder what an employed person is doing in *their* dumpster. Avoid this. Go for that look which says you're a clean, respectable person... but you might not be employed. This produces the "ninja effect." Not only will people ignore you, they will try to erase you from their memory. You can't ask for more than *that.*

The Wrong Clothing

My father used to drive a local taxi when he wanted to pick up some extra cash. My mother acquired a nice jacket for him to wear, perfect except for a small tear in the back. She fixed the tear and concealed the repaired area with a patch that she had created. The patch featured an image of a busy little taxi and the company's phone number. The owner of the cab company liked the patch so much that he paid my mom to create several for him.

One day my father was scavenging behind a local meat shop, picking up a week's worth of scraps for our dog. Somebody saw him. Well, about a month later we were having a discussion in my fifth grade class about all the poor, hungry people in our country. Tsk, Tsk. The teacher pointed out that even the employed, well-dressed people might be hungry, even if they didn't *look* hungry. That was when Suzie So-and-So spoke up and said her mother's friend's husband had seen a cab driver from "Busy" taxi company rummaging in the dumpster behind Joe Blow's Meat Shop. Suzie said her mother's friend's husband *knew* it was "Busy" taxi company, because of the distinctive jacket with a busy little cab.

I froze. I didn't say a word. My parents had taught me a long time ago to be careful with information, and Suzie So-and-So had no way of knowing that was *my dad*.

The teacher pointed out that, perhaps, the cab driver was picking up bones for his dog. (Smart teacher! I thought, but said nothing.) But, the teacher continued, people who couldn't afford dogfood were still poor, and wasn't that a shame? Tsk, tsk. The discussion meandered into the consumption of dogfood by little old ladies and other bullshit.

When I arrived home I reported the whole thing to my father in detail. He told me I had handled the whole thing exactly right and brought back "good intelligence." I swelled with pride. After that he made a point of wearing a more "obscure" jacket while dumpster diving. And, "retiring" from the job for the season, he gave the owner his jacket with the "busy little cab" logo.

So DON'T wear distinctive clothing with letters and logo.

Make sure your clothing is durable, won't snag easily, and cleans with minimal effort if soiled. Denim is excellent.

Mr. Clean — Dumpster Diver

By the way, you won't be washing your "diving clothes" every single day. People seem to think dumpster diving is dirtier and sloppier than it is. The majority of dumpsters are relatively clean and dry, except on the very bottom. People don't pour liquids in dumpsters, you know. At least liquids aren't discarded without a container. Even stuff like food waste is usually packed in plastic bags.

Now, dumpsters from eating establishments *are* messy. If you climb in or lean over the edge you will probably get grease on your clothing. I use a "dive stick" to avoid this. In any case, you won't be getting grubby very often. But accidents do happen, so wear something you can clean or easily replace. I haven't ruined any article of clothing for over a year, but it does happen. Once I spilled ink all over a pair of jeans which were a little "newer" than I would have liked. Thank goodness I didn't pay money for 'em.

Remember, most stains can be removed or hidden. This is particularly good to remember when you find clothing. People are fond of discarding stuff because of one little flaw.

By the way, *don't wear military articles of clothing*. Even though the color is great, even though the stuff is wondrously durable and washable, it is bad for two reasons:

1. Too distinctive. Too closely associated with "deranged" individuals. It makes you look dangerous and ruins that "ninja effect."

2. It provokes authority figures, especially if you leave the patches and insignias on the clothing. Most police have a military background, as well as plenty of mid-level managers with their anal retentive attitude problems. Seeing the words "U.S. ARMY" on some "dumpster diving punk" predisposes them to being a hard ass with you.

We'll discuss dealing with cops in more detail later. But as long as we're examining clothing, be conscious of your appearance from the viewpoint of an authority figure. Don't wear stuff with sports insignia, it makes you look like a "gang member." Don't wear flannel shirts with combat boots, it makes you look like a "skinhead" sympathizer. *Avoid any "look" dis-*

tinctive of skateboarders. Cops hate that shit. You may as well wear a "Death to the Pigs" T-shirt. Don't wear skull rings, "peace" earrings, ankle bracelets, "Harley" scarves, etc. In fact, if you are male avoid jewelry, especially earrings. If you have a "punk" haircut, wear a hood or a hat. Don't wear sunglasses, it makes you look like you're up to no good.

When you are *not* dumpster diving, feel free to dress distinctively, provoke authority figures, etc. I encourage you. But, while dumpster diving, cultivate that "obscure" look and you will do well. Don't worry about your "political statement." Dumpster diving is a hell of a statement.

Footwear

What you wear on your feet is extremely important. Consider the temperature and be prepared for rain. But most important is the *thickness of the sole.*

The area around dumpsters is frequently littered with glass. And dumpsters are full of broken glass — though not as much as most people believe. I prefer "shoe style" hiking footwear, though hiking boots which cover the shins will keep you safer. I once purchased a pair of black shoes with steel "safety" toes and thick, grease resistant soles. The shoes were ideal for my job at a printing plant but were also *perfect* for dumpster diving.

Standard gym shoes or casual footwear which encloses the foot will do if that's all you have at the moment, but be extra careful.

Gloves

Truthfully, I don't always wear gloves. You just don't enjoy the same sensitivity and dexterity you experience with bare flesh. Something comes between you and the dumpster experience, an artificial barrier. It's like showering with a raincoat.

I know I *should* be responsible and wear gloves. I mean, my God, there are AIDS infected junkie needles out there. (Though I've never seen a junkie's needle. My brother has found the actual junkies a few times.) I'll bet that I won't have a hard time convincing most novices to wear gloves while dumpster diving. You can always take the gloves off. Most important, it will help you avoid injuries from glass and keep your hands free of grease.

Leather or plastic works best, because you can wipe it clean.

Always wear gloves in subzero temperatures. Dumpsters are made of cold, cold metal and your hand can "stick." Ouch!

The Right Stuff — Gear

Dumpster diving doesn't require a great deal of specialized equipment, but there are a few items which help considerably.

Flashlight

Absolutely essential for diving at night. During the day you can usually see fine.

Don't skimp in this area. You can use a small flashlight, or a cheap one, but make sure the batteries are good. Nothing is more frustrating than shaking a dying flashlight, trying to make it cough up more light, when you *know* that dumpster is full of good stuff you can't see. Carry spare batteries as often as possible, and even a spare light is a good idea when you have a vehicle.

I use a "periscope" shaped camping flashlight, the kind used by the U.S. Army. I've dropped it, stepped on it, used it as a "hook" to pull stuff toward me. It just begs for more abuse. Use what's available, save your money, but when you do run out and buy a flashlight consider investing in something durable. I hate to sound like a commercial, but I think it's hard to beat those Army flashlights. They even have a red lens to reduce visibility, and a blue lens if your mood changes.

Rechargeable batteries used to save me lots of money, but recently I found that most of the batteries in discarded appliances are still "good." I have a drawer half full of batteries. I have a favorite battery, but for that type of endorsement I would certainly need some cash.

By the way, if you *find* a flashlight that doesn't work, try changing the bulb. Don't run out and *buy* a bulb, try the bulbs you have available first. Different flashlights take different bulbs, and the bulb type is stamped on the collar of the bulb. Most people will discard a ten dollar flashlight rather than purchase two bulbs for 99¢. And the batteries you find in the same box or bag will probably be good, too.

NEVER use matches or a lighter to illuminate a dumpster. Sure, you probably have enough common sense and manual dexterity to avoid torching the trash bin. Unfortunately, what you are doing with that match may not appear to be "rummaging" to paranoid passersby and burntout cops. Once I was hassled by a cop while I had a boxful of paperbacks sitting in the back of my pickup — he was more concerned with an outbreak of small fires in that neighborhood. If I had been using a match to see what I was doing, it would have been hard to convince that cop I was not a pyromaniac.

Unfortunately, trash piles and dumpsters are frequently the target of petty pyromaniacs, especially kids lacking proper upbringing. But the worst thing about matches is that *they just don't shed enough light.* And you really could cause a fire by accident. Don't even light a cigarette while you rummage. All the paranoid housewives will see is that flaring match. DON'T use a flame in any manner while dumpster diving. People throw away all kinds of flammable stuff. You could be standing in a puddle of turpentine, lighting your Lucky. Use a flashlight, and put that smoke off for a minute.

Dive Stick

A stick of some kind will save you a great deal of effort. You won't have to climb into the dumpster or lean into it if you can pull stuff toward you or push it apart for visual examination.

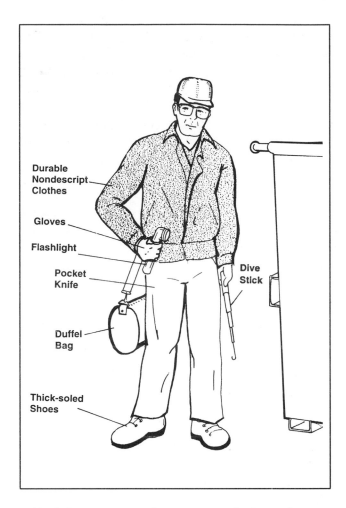

Durable Nondescript Clothes

Gloves

Flashlight

Pocket Knife

Duffel Bag

Thick-soled Shoes

Dive Stick

Feel free to use whatever works best for you. When I dumpster dive using a vehicle, I prefer a broom handle with a nail "hook" at the end, painted "garbage bag black" or wrapped with electrical tape. Don't put a great deal of effort or ornamentation into this thing, because if a policeman or store manager challenges you *the first thing you'll do is leave that stick in the dumpster.* You don't want to look like a "professional." More on this later.

When I *don't* have a vehicle, the last thing I want to use is a big, heavy stick. Distinctive. Dangerous looking. Some guy hanging around the dumpsters in the dark with a *big stick.* But you can get away with it using a vehicle.

As I mentioned earlier, I use an antenna from a portable C.B. radio. It extends six feet but collapses down to the length of a roll of Lifesavers. I've modified the end with a small metal hook. I can pull it out of my pocket, extend it and collapse it in the dumpster. *Nobody sees it.* Perfect. A small antenna will give you plenty of extra "reach" until you find a big antenna. (And these are easy to find.)

Remember, if somebody challenges you, quietly drop that "dive stick." Don't drop it with a suspicious "thunk." Knowing you may need to do this, keep a few in reserve.

My maternal grandmother, who took up dumpster diving late in life, used her cane in this manner. But she could walk around with a cane because she was a little old lady.

Loot Bag

I use a gray Adidas bag. The "Adidas" is almost worn off the bag. A bag of some type is valuable for carrying off your finds if you are on foot. Sure, you can use the boxes in the dumpster. But if you find something good in a rain soaked cardboard box, what can you use? Besides, walking away from a dumpster with a gray book bag is less suspicious than walking away from a dumpster with a handful of magazines, a metal Christmas tree stand, etc. Police can't search the bag without your consent. (Theoretically. They aren't called "pigs" because they like to eat a lot.) Gym bags work well for this. Small duffel bags are good. Avoid distinctive, readable logos.

Bag Blades

A person can make a lot of great finds without ripping or cutting open trash bags. But if you want those extra good, hidden items, *you gotta slice open some trash bags.* We'll look at the best way to accomplish this later, but right now let's consider the equipment.

Think straight razor. You want something that can slice one millimeter of plastic. You don't need to gut a deer. Get something really sharp, not thick but not flexible.

I prefer to use a small pocketknife with an extremely sharp blade. A razor works better, but I don't relish explaining to a cop why, exactly, I'm carrying a straight razor in a dark alley. *Know local laws.* Over a certain length a knife is a "concealed weapon." Remember, this is another item you must be prepared to abandon, so go the cheap route.

Weapons

In two decades of dumpster diving I have never been attacked by a stray dog, a wino, a junkie, or anyone other than my dear baby brother... but I still carry a chemical repellent. That's for stray dogs. I'll try my "bag blade," flashlight and boots on any human attackers. When I use a vehicle I carry a firearm, carefully concealed. But that has nothing to do with dumpster diving, just my general perception of local conditions.

In my opinion, a dumpster diver needs less protection than a vacuum cleaner salesman. After all, who thinks a dumpster diver is worth robbing? And most of the psychopaths who attack vagrants only pick the drunk, helpless ones. My greatest worry is that I'll be arrested for violating some bullshit local ordinance and the arresting officer will find my weapon.

Self-defense is a personal issue. Do what feels right for you. But if you carry a weapon while dumpster diving you should carry one all the time. Don't pack a pistol because you're involved in this "dangerous" dumpster diving. *Rummaging in dumpsters is no big deal.*

Vehicles

A vehicle is an extremely valuable asset for reaching primo dumpsters and hauling off large items and/or large amounts. But don't overuse your vehicle. If you are just checking out your block, and you're in good health, *walk.* Vehicles have license plates and mark you as a "professional." Besides, walking is good for you.

A bike with a large basket is very valuable, extending your range and saving effort.

"Hardcore" dumpster diving certainly requires a vehicle. Trucks work best, but cars will do fine if that's all that's available. A small truck with a detachable "topper" is the perfect vehicle. A van or an "El Camino" works well, too.

Keep a few empty boxes in the back of your truck. Not only are boxes handy while making the haul, but they provide an alibi. The number one dumpster diving excuse is, "We are just looking for boxes." The boxes can also conceal your loot. A few plastic bags are handy, too.

Back to the vehicle itself. Avoid distinctive stuff on your vehicle like political bumper stickers, Virgin Mother statues, yellow ribbons on the radio antenna, etc. "Dive sticks" should be kept out of sight behind the seat or in a closed truck bed compartment.

Don't bother trying to "shine your headlights" into the diving area. All you will see are long, angular shadows and the blinding headlights. You'll also light yourself up like a casino, attracting attention from cops and passersby.

Bungee cords, ropes, tarps, canvas, etc. are valuable for securing big loads, like furniture during campus moving day. Keep a piece of RED CLOTH handy if you haul off long planks which extend past the end of your truckbed. Rope is good for securing an overstuffed car trunk, also.

Keep your vehicle in good repair so you don't get stranded behind a store with a load of discarded produce. Drive safely. Don't block your rear view with a load of stuff. You may drive just fine, but if a cop pulls you over he may figure out your game and decide to be an asshole. The line between a policeman enforcing useless, overly-intrusive laws and an asshole is a fine line, indeed. BEWARE OF PAPER FLYING OUT OF YOUR VEHICLE. Secure that stuff, or you're asking for trouble.

Little Extras

Lots of things are not necessities, but make dumpster diving more fun and pleasant. A Thermos with hot drinks, for example, is really great in the winter. I like to wear a Walkman while strolling from one dumpster to the next. I don't wear it while rummaging, however. If somebody is walking up while my head is in the bin, I want to hear them approaching.

A First Aid Kit is a smart item to carry in your vehicle. I found a nice first aid kit in a dumpster, cleaned it up and restocked it with sterile gauze from my hospital. Some months ago I treated a badly injured motorist. He probably didn't care where I acquired those smelling salts and road flares. Suntan lotion should be worn to avoid sunburn.

Don't pack for a safari, but feel free to have fun. If you're a real novice at this sort of thing, just bring the basics at first. Once you get a feel for dumpster diving, branch out a little and bring more extras.

A notebook to record your finds *after you get home* is invaluable. A personal computer could be valuable, too. Take a picture of yourself and send me a copy — don't let some old lady see you doing it!

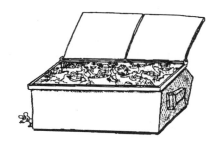

TWISTED IMAGE by Ace Backwords ©1993

Chapter 6
Diving Techniques Illustrated

Over the course of two decades of dumpster diving, I have never been arrested or hurt seriously. However, I've been hassled by police, sustained several small injuries, and grabbed a few things that I wished to God I had not. I've missed some great opportunities, by being ill prepared.

Fortunately, I've learned from my mistakes and carefully examined the mistakes of others. There's no reason you should have to endure these unpleasantries if you will permit me to show you the way.

Dumpster diving isn't hard and it isn't dangerous... most of the time. I wouldn't want you to think you're going to be dealing with "broken glass booby traps" or "aerosol land mines" frequently. So don't act like you're working for a bomb squad. Just relax. Have fun. But keep safety in mind.

Here are some "hard won" tips on keeping safe, having fun, and maximizing your "haul" without wasting time.

LOOK, LISTEN, SMELL, FEEL. Use common sense and your other senses.

VISUALLY INSPECT the dumpster. Quickly get an idea what's in there and how you're going to approach things. (Or NOT approach. Most dumpsters are "dry wells." If you want a good haul, you have to hit a lot of dumpsters.) If it is night, shine your light around, *being careful to keep it below the rim of the dumpster*. This will reduce your visibility to passersby, including cops, dramatically.

Look for broken glass, gooey messes, heavy articles that could slip around and crush your feet, critters, anything you'd like to avoid. Look for indications of "good stuff," such as boxes sealed with tape, intact food containers such as cans and boxes, and other indicators we'll discuss later.

LISTEN. Listen for skittering noises, but don't let the gentle rustle of plastic or your own imagination make you believe you just stumbled on the secret lair of King Rat. In tens of thousands of dives I have seen *one rat*. That's right, *just one*. Now, I dive some pretty nice neighborhoods where good sanitation is practiced. There are more rats in filthy, neglected areas. But the common assumption that dumpsters are full of rats *just ain't so*.

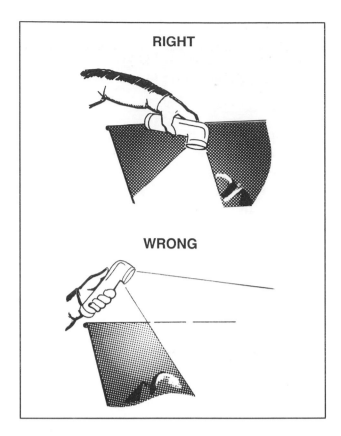

RIGHT

WRONG

Always keep your flashlight below the dumpster rim.
We don't want to disturb the neighbors.

Yes, there are flies, but only in refuse containers where food is discarded *frequently*, or where food refuse *remains for a long time*. In other words, food refuse may be discarded in a residential dumpster but if trash pickup is frequent the flies won't accumulate. Bees and hornets love fruit refuse and sweet, sticky stuff. Exercise care if you hear the little darlings. Cockroaches? Practically non-existent. They like it under the sink of a clean kitchen better. The idea that a dumpster is full of roaches, maggots, rats and mice does not hold up upon closer examination.

Behold! The modern dumpster is designed for the purpose of *keeping vermin out*. With its lids, vertical steel sides and smooth, featureless surface it does an excellent job. The dumpster where I saw the rat had been damaged by a vehicle, allowing the rodent access. Ironically, it was a bookstore dumpster filled with paper. The only food in the dumpster was a half-eaten burrito from some employee's lunch. The poor,

starving rodent ran past my hand, terrified, and leaped off the rim of the dumpster. I had a bit of a complex for about a month, but survived the experience and learned something about damaged dumpsters.

You could encounter rodents in the food refuse littering the area *around* dumpsters, but rats and mice are a pretty meek lot. Don't worry about them. Besides, business and property owners frequently use poisons to keep their dumpster area rodent free. Like paranoid idiots, they pour the majority of the poison into the dumpster and sprinkle a little bit on the ground. What a waste of perfectly good poison. No wonder it takes more to kill the little darlings every year.

Be aware of this if you see powder dumped liberally over the contents of your favorite trash bin. I'll discuss my encounters with the stuff later.

Now, CATS are a different matter. I've had plenty of cats come flying out of dumpsters, eyes aglow, screeching, looking for all the world like King Rat. I've never been bitten or attacked, however. I happen to like cats, so I just laugh the whole thing off and enjoy the "cheap thrill."

A few times I have found sleeping vagrants. I would hate to see anyone crushed in a garbage truck, something that happens from time to time, so I take the initiative of calling the police. My little brother, Slash, likes to rap them on the head, flash one of his pseudo-official IDs, and tell them he just implanted a microphone in their skull. I don't recommend this kind of behavior.

Anyway, LISTEN, but don't let your imagination play games with you. Rats and mice in dumpsters are a rare event.

SMELL has never helped me to locate anything good, though it has helped me to avoid a lot of dirty diapers and bags of discarded kitty litter. (It has a powerful ammonia smell when disturbed — I don't know how cats use the stuff.) People like to say smells are "good," or "bad," but really these are just smells.

My little brother has tried to use smell in an interesting way. When he was ten years old he found a live puppy in a residential dumpster. Apparently, somebody thought this method of puppy abandonment was more innovative and humane than dumping the dog by the side of the road. Anyway, Slash adopted the dog and tried to teach him to locate food in dumpsters. He figured the dog would be predisposed to this sort of thing, since it had early experience with dumpsters.

It never worked, however, despite Jed's frequent claims of success. The problem is that a dumpster is full of interesting smells. Dogs don't find these smells offensive, but *fascinating.* Put them in a dumpster and they go nuts. They can't differentiate; given a choice between an old, rancid piece of bologna and half a dozen cartons of milk, they take the bologna every time. If a dog could be taught to seek *one thing,* like cheese, it might work. But what would be the point? It's not like trying to locate cocaine in luggage. You are free to tear everything in a dumpster apart and look inside. Besides, the dog could get cut by glass. Based upon experience, I would say DON'T BRING YOUR DOG. It would be sheer animal abuse to keep a dog locked in the truck while he watches YOU rummage in dumpsters.

Use your sense of TOUCH with your other senses. Poke a trash bag with your hands and you will be able to determine rapidly if it is full of paper, food refuse, clothing or miscellaneous items. And, by the way, this becomes easier with practice. Just as you develop "diver's eyes," you'll develop "dumpster fingers." No, this isn't a fungus. Dumpster fingers means you can poke a trash bag, determine it is filled with pizza boxes, and know immediately *what brand.*

But be careful that you don't jump to conclusions all the time. TOUCH is a good way to avoid opening bags that are full of slop, but it's no substitute for slicing the bag open and having a look.

Remember, these aren't things that require a great deal of time or thought. Open the lid and start poking around, using all your senses at once. Every dumpster is delightfully different and full of surprises.

#1 Problem — Glass!

One piece of advice will keep you safe ninety percent of the time: WATCH OUT FOR GLASS!

Assume every dumpster contains razor sharp pieces of glass. Even if it *seems* the whole dumpster is full of violated, coverless paperbacks and bubble wrap, all it takes is one "Tropicana" drink bottle to slice you. Don't forget, people use commercial dumpsters to discard trash from their vehicles, and every place has employees eating lunch. Even a paper recycling bin will often contain broken bottles. People throwing stuff away do funny things. They might throw a broken mirror in a box and toss a nice pile of clothes on top. Don't be paranoid, just be careful.

The area around a dumpster is often littered with small shards of glass. Don't sit next to a dumpster while rummaging through a box — squat! Don't place your bare hand on the ground to push yourself up. Be careful of glass, and most of the other stuff will take care of itself.

The Dumpster — A Wonder to Behold

The modern dumpster keeps garbage in, vermin out. It has incredible capacity, which just encourages people to fill it with stuff.

Dealing with dumpsters isn't dangerous if you simply use common sense. Here are a few things to approach with care.

LOOK OUT for those heavy metal lids suspended in the "up" position. NEVER drape yourself over the edge of a dumpster with the heavy lid hovering over you. I've seen the wind buffet those things and send them crashing down suddenly. Informal experiments with plastic pipe have lead me to conclude you could break your hand or even your *neck* in this manner. Use a dive stick, keeping one hand on the lid so you know *exactly* where it is. If you must get inside the dumpster, let the lid down or stay below the rim of the dumpster, keeping your hand on the lid so you know where it is. Having a partner is handy in these situations.

"Two-holers" are frequently secured, but access is allowed via the small doors.

LOOK OUT when lifting lids so that you don't get your hand pinched. Remember, those lids are abused by sanitation workers and are *frequently off center*. In other words, if you lift one lid the other lid would be off center and it will rise, too. However, when you least expect it the other lid will crash down very noisily — unless your hand is under it. If you lift one lid and the other rises, put down the first lid and start with the other. Usually it will rise by itself.

BE QUIET with lids. You don't want a noise complaint. Toss a lid haphazardly and it can be heard for blocks. However, don't be paranoid. People make noise all the time. Citizens aren't constantly poised over their phones, ready to call the cops. So go ahead and toss those boxes around or talk quietly to your partner. *Nobody is going to give a damn.* If you do bang a lid noisily,

don't take off like a scared rabbit. It's no big deal — most of the time.

Beware of dumpsters sitting off-center on platforms — they can smash your feet!

Watch out for heavy lids suspended in the upright position — they can break your neck!

WATCH OUT for dumpsters that sit on raised platforms. The dumpster can tip off the edge of the platform while you lean into it. Make sure it is steady. Watch out for dumpsters missing wheels, too. You can crush your foot leaning into a "three-legged" dumpster.

Watch out for sharp metal edges. Oh, and watch out for rainwater collected on dumpster lids. It's not dangerous, just cold and wet. Water can collect *inside* some of these hollow plastic lids once they are punctured, which happens frequently. Again, don't be paranoid — just be careful.

The classic dumpster pose. Position dumpster below navel, not in your gut. Lean into that sucker!

Exploring Dumpster Depths

Don't get inside a dumpster unless you can't avoid it. First, it draws a lot of attention to you. If somebody walks to the dumpster to deposit something you can't just walk away and do the "ninja" thing. As much as possible, *lean* into the dumpster or use a dive stick. Secondly, getting inside a dumpster means a higher level of risk. You may find yourself in the same metal box with a terrified cat. Or, far more likely, you could step on some glass.

There's a certain trick to leaning inside a dumpster. I call it the "see-saw."

Most beginners are afraid they'll fall inside, so they lean in a very conservative manner. They position the rim just below their ribs or right in their diaphragm. This is extremely uncomfortable and makes breathing difficult.

The "trick" to comfortable diving is to really lean into that sucker. Position the edge of the dumpster below your navel, so you are free to breathe. Then simply use your legs to balance. If you want to lean forward, raise your legs. If you want to get out again, drop your legs and arch your back. If you find something heavy in that dumpster you may not be able to ease it over the edge. Instead, you'll have to hang on to the object and "ooze" back out. You'll have to do the same for fragile objects, like boxes of glassware. But this method allows you to toss out dozens of light, non-fragile items without the need to reposition yourself. But you do need to practice a bit, or you'll fall into a dark, uncharted dumpster head first. Keep your balance — DON'T DRINK AND DIVE.

You can "tunnel" your way through empty boxes in this manner, looking for buried goodies. Finding stuff at the bottom with only your flashlight to illuminate things, hanging in space while listening for somebody to approach is a rush. Remember when you do the "see-saw," that stuff in the dumpster can be used to maintain your balance or leverage your way out.

You might note that this position looks rather like a person diving into water. I certainly didn't invent the phrase "dumpster diving," though I'm doing my bit to popularize it. The phrase probably came about because of this useful position, the "classic pose" of a professional dumpster diver. Don't do it unless you are reasonably agile. Don't do it if you bruise or break bones easily.

By the way, the edge of a dumpster can be very uncomfortable, as well as dirty. Feel free to grab a piece of cardboard to use as a "cushion."

This is especially useful on the dumpsters with thin rims, thin being less than an inch.

That's not for lifting the dumpster — it's a foothold.

Note also that dumpsters have convenient "footholds" which allow you elevated access. This is even more tricky than the "diving swan" position. One stiff breeze and you can topple over into an uncharted dumpster. I prefer to put my knee on those footholds rather than stand on them, particularly since the foothold can be wet or coated with grease. Kids, with their small feet and shorter height, can often stand on those little ledges. But the proper place for a child is outside a dumpster, picking up articles and keeping a look out. We'll discuss kids more at a later point.

Of course, there are plenty of times you need to climb in the old bin. This may be necessary with extra large dumpsters, or to secure articles which are not boxed or bagged. For example, if you see a box of cracked coconuts but the box has decomposed in heavy rain, you may need to get inside and find a nice waxed paper box or a peach crate. You could probably do this in the "swan" position, but that's a lot of blood rushing to your head. More importantly, it would take

longer and increase your odds of being really noticeable — more than a brief climb in the bin.

It's no big deal. Carefully climb in. Despite the title of this book, never dive, leap, jump or bound into a dumpster. Always assume there is broken glass. Grab ahold of the sides of the dumpster and carefully lower yourself.

Don't step where you can't see. If it is night, carefully examine the area where you are going to stand. Don't try to lower yourself and hold the flashlight at the same time — it won't work.

Use a piece of cardboard when leaning over a thin or dirty dumpster edge.

Step on a piece of cardboard, a board, or whatever relatively flat surface you spot. If you are going to stand on the bare bottom of the dumpster, toss a piece of cardboard or something on it. Even standing on a sour-smelling milk carton or a sauce-splattered pizza box is better than standing on the bare floor of a dumpster. Like the floors of movie theaters, the bottoms of dumpsters are coated with a mysterious sticky substance. The stuff is not dangerous, merely unpleasant and hard to scrape off your shoes — like fruit scented tar mixed with sour milk. I've seen exactly the same stuff on the floor of neglected walk-in freezers, so I conclude it's composed mostly of fruit juice, soda pop, grease and other sticky stuff. Fortunately, dumpsters are seldom empty for long and you will rarely confront the sticky bottom of a trash bin.

So step on something reasonably flat. LISTEN as you step, so you can determine if there is broken glass under your feet. Even if there is broken glass beneath your feet, you probably won't have a problem if you step slowly and carefully. If you seek a better place to stand, grab the sides of the dumpster again so you can ease yourself. Don't commit all your weight at once. Whatever you do, don't try to stomp down a protruding surface. It could be the broken neck of a gallon cider jug, a discarded hoe, anything.

Don't walk around atop a pile of refuse in a big dumpster. Peer inside and use a dive stick.

Once you find a good place to stand, stay there and move things around with your hands. Don't prance around inside a dumpster. And beware of freshly fallen snow. Most people won't step directly on a broken bottle, but plenty of dumpster divers will step on stuff they can't see if a fresh coat of snow covers it. Snow won't

protect you from shards of glass. Use your dive stick or whatever you can grab to clear snow off the area where you intend to stand. Don't be paranoid, but be careful.

THREE THINGS TO WATCH OUT FOR, based on my own hard lessons and those of fellow divers.

WATCH OUT if you feel a funny "pressure" or springiness or sense of tension in your foot. You could be standing on the thin slats of a wooden crate, a dresser with a cheap fiberboard back, a couple panes of plate glass suspended on crates, all kinds of stuff that can give away suddenly and cut you, to boot. You could also be standing on something sharp and protruding, ready to stab through the surface you're standing on. When standing in a dumpster, never lose track of what your feet are telling you.

I've never been seriously hurt in this manner, though I've received nicks and scratches. Usually I receive the scratches trying to step back out again and getting caught up on jagged edges. So if you feel your foot crash down suddenly, don't yank your foot out reflexively. Raise the foot carefully, so you don't cut yourself on the edges of whatever you just smashed through. For god's sake, don't walk around on ten foot mounds of garbage in those tall, super-dooper dumpsters. That mountain of bags can give way and hurl you into a pile of broken glass. Use a dive stick, and enter a tall, over-stuffed dumpster only after carefully assessing the situation and exhausting other options.

SECONDLY, BUT WORST OF ALL, are those damned aerosol cans. I've never been injured by one, but on a few occasions those things have scared the hell out of me. If you are standing on top of one it usually won't go off — though I certainly don't recommend testing your luck. By walking around, however, you can jam some object like a board with a protruding nail into one, setting off the pressurized container with a bang or a scary pssst! or a variety of other noises. Once a can of FDS spray went off under my feet with a loud FUDST! It was, I thought, kind of like Garp's father in *The World According to Garp*, saying GARP! right at the end of his life.

In any case, most stuff in a dumpster isn't explosive, and even pressurized cans are more of an annoyance than a danger. I wouldn't want pieces of metal in my eyes, however, so avoid stepping on these things or crushing them accidentally.

LAST, BUT NOT LEAST, watch out for heavy articles. You can shift something in a dumpster and send a discarded stereo speaker toppling off a mountain of plastic bags right on your foot. Watch out for heavy stuff suspended in a manner which tempts gravity. This is the kind of thing that happens more frequently when two divers are working a dumpster. Be careful your partner doesn't shift something unexpectedly, or that you don't topple something on your comrade. In most cases it is more efficient to have one person inside the dumpster and the other person loading salvaged articles into the vehicle. Have the more experienced person in the dumpster, unless that person is physically frail or disabled. In that case, have the stronger person in the dumpster and let the more experienced person give verbal guidance. My brother, sister and I were frequently guided in this manner by my disabled father, and it worked well. But don't toss a kid in the dumpster just because you are lazy. They won't be able to quickly calculate what's valuable. More on kids later.

In my experience, the majority of accidents happen when people are excited over a "bonanza." They find a month's supply of canned veggies, get careless, throw things around, step without looking, etc. That's when people get loud and call attention to themselves, also. When you find something great and you just can't believe it, indulge yourself with a quiet "YIKES!" or something. But DON'T yell, DON'T get careless.

Loosen Up — Have Fun

There's nothing worse than some survivalist wannabe who treats his fellow divers (especially kids) like he's a squad leader wading around Nam instead of a balding guy rummaging for "mystery cans" behind SuperSaver. Feel free to make jokes, play games, and so forth. Some of the best times I've had while dumpster diving were with my little brother, "Slash." He pretends to be everything from an old lady to a ninja, a pirate, a paranoid guy with delusions of "secret service" grandeur, even a naive exchange student from New Zealand.

Yes, kids do need supervision and guidance or they'll get hurt. But twelve-year-olds and teenagers show a great deal of common sense if you let them. After the age of twelve I showed my parents a trick or two in the ol' bins. Besides that, teenagers frequently have more energy and more time on their hands than adults. After the age of sixteen I was doing "the route" without my parents, accompanied by Jed, who was thirteen. My mother and father helped us "process" and sell the dumpster goodies. They would praise me, Jed and Bekka extravagantly. We were only kids and yet we were truly "providing for the family." And we weren't doing it by working some slave wage job, either. We were doing it on our own terms, at times we found convenient, and we were having fun.

Methods

Looking through a dumpster for useful articles isn't tough, but there are ways to maximize your take and minimize your time and effort.

First, go right for the good stuff. If you look inside a big trash bin and one corner appears promising, start at that corner. Don't save it for later — you may have to leave suddenly. The faster you remove the "good stuff" from a dumpster, the more room you'll have to move around. Also, removing one piece of loot will frequently uncover a pocket of good stuff. It's kind of like mining. You don't methodically tear the whole mountain apart. Rather, you seek a "vein" and dig until the vein is played out.

Certain things are indicators of goodies. For example, boxes sealed with tape are almost always filled with good stuff. If it isn't good stuff, it is seldom garbage. You're more likely to find somebody's obscure documents than, say, empty bean cans. ALWAYS grab sealed boxes behind bookstores. These boxes contain books and magazines with their front covers ripped off. Toss those sealed boxes on the back of the vehicle and keep diving. In fact, when you find ANY box that seems to be crammed with goodies, or even really promising, just grab it. You can check the boxes later, away from the dumpster.

Produce boxes are always good to check in supermarket dumpsters. In fact, check any sturdy-looking box. Employees will always grab a midsized, relatively tough box when throwing out, say, a whole freezer full of "expired" t.v. dinners. I have also noted bakeries where plastic-lined fifty pound flour sacks were used to discard "two day old" bakery items, still conveniently packed in plastic wrapping and styrofoam trays. The problem isn't finding them — it's using them up.

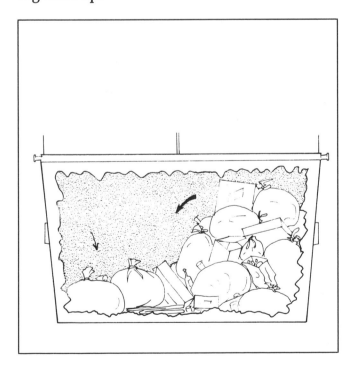

Check the stuff on the low side first,
then work your way down the high side.

Always check inside unusual containers like discarded suitcases, duffel bags, purses, garment bags, trunks, dressers and so forth. These items are frequently full of good quality discards, even if the container is in bad shape.

Rip or slice the lining out of purses and suitcases which are too beat up for use. Sometimes you will find small, valuable items lost by the original owner. I've found several Mercury dimes in this manner. Check stuff left beside the dumpster, first. These boxes and bags frequently contain high quality items that people feel guilty about discarding. They want you to pick it up. Don't say to yourself, "They can't be throwing *this* away." They are.

So check those "hot spots" first. After that, check the dumpster methodically. If the stuff in the dumpster is piled mostly to one side, check the low side first. When you are done with the low side, you can move articles from the high side to the lower area, maximizing your search. Carefully check under *both* lids. Check *both* sides. An area may look unpromising, but viewed from a different angle hidden goodies can be spotted.

You may have to temporarily remove light, bulky articles while you search. For example, a Christmas tree or an empty television box can really get in your way. Pull the item out and check the dumpster, then toss the item back. If I know I'm going to check the dumpster the next day, too, I'll sometimes tear a bulky box apart and toss it back inside. This creates a nice layer which will be easier to check the next day.

Learning how much time and effort to spend on individual dumpsters comes with practice. Generally, I will carefully check my "hottest," most dependable dumpsters and merely poke around in others. I always take more time and care with grocery and bookstore dumpsters, because finding even a few good articles justifies my efforts. In residential areas I poke around, slice a few bags that look promising, then move on. I won't spend a great deal of time and effort on these dumpsters because they don't pay off as frequently as bakeries, bookstores, grocery stores, etc. All the same, you won't find very much unless you make the effort to toss a few boxes around, slice some bags, etc. A dumpster which looks dreary and unpromising can quickly become "hot" if you shift one or two boxes. And vice versa.

When I'm checking out a residential area that I'm considering incorporating into my route, I'll merely skim at first. It gives the people in that area a little time to become immune to my presence. After that I rummage intensely. But when I'm checking out a new commercial dumpster I really check it out good, so I can quickly develop an idea how "hot" it is. If a week goes by and it's never hot, I'll just skim it after that. Remember, these "heat patterns" change constantly. A shake-up in management can lead to new waste patterns.

During moving day in campus areas I will check out strange dumpster territory with methodical care. Why? Because it pays off. I always make an effort to focus on the biggest, most dependable pay offs. If I can't check all my dumpsters I'll check my best ones. Every dumpster is different and you learn with experience.

Bag Blades

Using a bag blade will help you increase your take. The trick is to figure out which bags are promising before you waste effort slicing the bags open.

I love white trash bags because you can see through them slightly. Press the surface of the bag closer to the contents and you'll be able to see more. If a bag has interesting angles, if it's really heavy, if it rattles or plays a few bars of "We're In the Money," slice that sucker. Watch out if it sounds "squishy."

When you slice a bag, hold it up and cut the bottom off with a couple circular swipes. An expert can "spin" the bag with one hand, holding the blade steady, and take the bottom off in one quick motion. All the contents will fall out the bottom for examination. WATCH OUT for your feet. WATCH OUT for clouds of choking dust from, say, discarded vacuum cleaner bags. Keep the bag close to the surface of the other trash, to minimize damage to fragile items as the contents fall out.

IF a bag is caught on something or wedged tightly, IF you can't lift it, IF you're 90% sure it's full of good stuff but you want to check, then cut off the top of the bag like a coconut cup. Otherwise, slicing off the bottom works best and will save you effort.

Even if you don't have a bag blade, you can easily pull a plastic trash bag apart with your hands. Poking a hole in it first works well. Trash bags are NOT tough, or even "hefty," despite the commercials.

But, for God's sake, be careful with that bag blade. Don't cut yourself or your partner.

Dumpster Etiquette

This ain't Emily Post here. The point is to stay out of trouble. And ninety percent of "dumpster etiquette" can be summarized as follows: DON'T MAKE A BIG MESS!

If a lid is closed when you arrive, leave it closed when you leave. If it's open, leave it open, unless stuff is blowing out. Avoid slicing bags outside dumpsters. If you *must*, cut off the top, not the bottom. Toss every scrap you don't salvage back in the dumpster. However, if a few scraps get away from you, don't fret. Just be reasonably careful. Throw a few unopened bags, large boxes, or other articles on the loose refuse to prevent it from blowing out or rousing unnecessary suspicion. If you remove a large article from the dumpster to obtain more room to rummage, put it back, of course. These actions will keep residents and employees from becoming enraged. Don't make a mess and they usually won't harass you or call the police. Throw crap all over and they certainly will. Often I leave a dumpster area *cleaner* than I found it.

Another major thing to remember is AVOID PEOPLE AND THEY WILL CAREFULLY AVOID YOU. If you are rummaging and from the corner of your eye you spot somebody coming to "make a deposit," grab your things and leave. Don't run like a criminal, just proceed to leave like you're finished, even if you just lifted the lid. If you remain, some sort of verbal contact is hard to avoid. Most people can live with the idea somebody is picking through their garbage, but they don't want to deposit the refuse into your eager hands. After all, if I notice a copy of *Hot Babes Over Fifty* in a dumpster, it's no big deal. But if somebody is discarding their intensely personal trash and I'm right there, ready to dig through it, it becomes embarrassing to the person doing the discarding. Embarrassment turns to anger pretty quickly, and next thing you know they're telling some cop bullshit about the sanctity of their refuse. Like *they* needed it. Like people at the landfill don't see the stuff. Idiots. Anyway, simply leave and return later. This will prevent a lot of hassle.

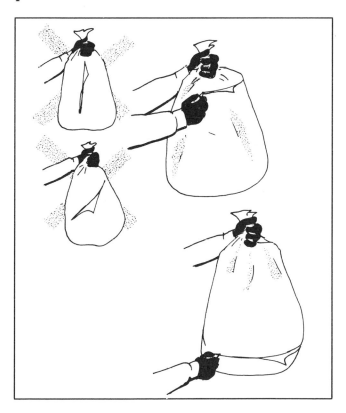

The best way to slice open a garbage bag is to lift from the top, cut from the bottom.

If somebody comes upon you unexpectedly, make every effort to ignore them and avoid contact. Ignore them even if they speak to you. Act deaf. Bend over and act intensely interested in some box or bag. Don't say anything witty to diffuse the tension, *just ignore them completely.*

They will probably ignore you and then try earnestly to forget that you exist.

Slash makes a point of mumbling things about "CIA plot to monitor my ejaculations," but then Slash relishes confrontation with authority. He once spent 45 days in jail for refusing to stand for the judge in traffic court. Anyway, avoid contact and save yourself a lot of hassles — unless you're into that sort of thing.

To a certain extent, the same rules apply to commercial dumpsters. However, stock boys discarding crates of bruised peaches don't really give a damn about dumpster divers. They're not throwing away something personal. But avoid contact, anyway. If the stock boys discard stuff right in your face and completely ignore you, *just* ignore them and keep rummaging. After a while, a few of them might say, "Hey, buddy, here's some good bread." Thank them, smile broadly, but *don't* converse with them. They may figure out you're not poor, dumb and desperate. You may get them in trouble with their boss, also, who is usually an asshole. BEWARE of a possible attempt to poison you, but don't be paranoid. There's a lot of sick people out there, and somebody who appears to be a transient makes a tempting target. That's why I prefer the look which says, "I have a home, somewhere — but maybe not a job."

Bleeding hearts may sometimes make an effort to speak with you, offering some sort of help. Play this by ear. Slash once obtained $20 in this manner to "buy medicine for my little brother." The young lady who gave him the money was wearing a cotton *serape* and a button which said, "I don't eat my friends." Slash ran out and bought bullets for his squirrel rifle. I think this sort of thing is dangerous in the long run. If they *offer* help, *take it*, but don't make up elaborate stories unless you have an independently confirmed knack for it. DON'T give them information about yourself. They may not be as sympathetic as they appear. WHATEVER YOU DO, don't speak to the news media. This just stirs up anti-dumpster diving programs. The people who are sympathetic to "vagrants" don't, as a rule, have very good garbage.

Certain dumpsters must be treated almost like burning buildings. Bookstores would be the best example. These people rip off the front covers of magazines and paperbacks, sending these back to their distributors for rebates. They become very irate if you retrieve these books and sell or barter them. Best dive these dumpsters at night or *like lightning* during the day. If I have to dive a dumpster like this during the day I send my partner on foot to "scout" the dumpster, quickly lifting the lid and peering inside. If sealed boxes or scattered paperbacks are noted we dive the dumpster in a rapid manner. I back the truck up with my partner in the truckbed. My partner lifts the lid, gets inside *carefully*, and rapidly, calmly hurls the boxes on the truck. He climbs out, closes the lid and sits down. I throw the truck in gear and take off. The whole time, I've been keeping a look out. Diving in this manner we have obtained over a thousand paperbacks in *ninety seconds.*

DON'T use these kind of extreme methods merely to avoid embarrassment. This method should be used only in a maximum profit, high risk situation. For example, you KNOW the dumpster is full of food but the asshole junior manager has a habit of walking out and saying "Get da hell outta der." Sometimes we obscure our license plates with mud for good measure, cleaning it when we're out of sight. DON'T drive around with muddy plates while using a vehicle, or you'll have problems.

Back to the bookstore dumpster. Once, Slash noted a coupon for two free packs of cigarettes in one issue of a "celebrities and events" magazine. We had obtained approximately fifty copies of that magazine. Slash carefully tore out each and every coupon, selling these to his friends for a buck apiece. He used this cash windfall to purchase a used pistol. (What can I say — the guy loves weapons.)

First "Solo" Dive

My first "solo" dive was a bookstore dumpster. It was a memorable experience.

When I was twelve, I started riding my "almost new" ten speed to town where I would hang out all day — mostly at the library. Dumpster diving on my own wasn't something I did immediately. After all, my parents had all the good spots covered and I certainly didn't lack anything. But it was I who "discovered" bookstore dumpsters for my family. Prior to that we obtained books in residential areas, trading titles we didn't want for other titles at used bookstores.

I knew "solo" dives were bound to cause conflicts with authority if I was caught doing it. People won't confront an adult but a kid dumpster diving provokes more comment. Where are his parents? Who is allowing this behavior?

The first solo dive was kind of a lucky accident. The dumpster was neatly identified with the name of the bookstore. I *love* this kind of convenient labeling. It *must* be for my benefit, since the sanitation workers don't give a damn. Anyway, I *loved* that bookstore. I read most of the novels of Robert A. Heinlein there. I would go in there and read all day. So that was why I peeked in the dumpster. I knew the store well and loved their merchandise.

The first thing I spotted was a *Star Trek* photo novel. Oddly enough, it still had its covers. You will periodically find books with covers, mostly "classics" with a 50% OFF! sticker. Bookstores seem to think it's better to discard these books than sell below a certain price. Most of the famous titles I've read in my lifetime were books obtained in this manner.

Anyway, my mind fairly reeled with the possibilities when I spotted that *Star Trek* book. I liked *Star Trek,* but the Kietzer twins were nuts about the old series. I could read this book, enjoy it, then trade it to the twins for comics. I'm a bibliophile, and right then I was looking at a biblio *pile.*

Only one problem: this was daytime. My parents knew I was a good kid, but I still had to be home before dark. And this dumpster wasn't in an alley, but a busy parking lot outside a midsized shopping mall. Visibility was a big problem.

I locked my bike up in front of the mall, then walked around the back. When nobody appeared to be looking I climbed in the dumpster and closed the lid. I gathered up all the books that caught my fancy, carefully and quietly, putting them in a box. Then I peered outside, barely lifting the lid.

Damn! The parking lot had grown busy while I rummaged. This was going to be tricky.

I continued to watch, waiting for a moment when I could pop out. Nowadays I'd just climb out, and to hell with bystanders. But I was only twelve, diving solo for the first time. My hands were full of literary loot, and I was overly paranoid. I waited and waited.

I heard staccato footsteps. Without moving my head more than a few inches, I peered off to the side.

It was a saleslady from the bookstore. This was the lady who always made a point of saying, "May I help you to find something?" She would do this even when it was clear I was progressing in my copy of *The Puppetmasters* quite well, thank you. Oh, excuse me, *their* copy.

"Sheeez!' I gasped, seeing she was walking right toward my hiding place.

I sat down slowly, quietly. I picked up a box with shaking hands and put it over my head and shoulders, pulling my knees close to my chest. My hands covered my chattering teeth.

I was really scared, expecting arrest and a severe bawling out. My heart slammed in my constricted throat, but I didn't move a muscle.

I heard the lid squeak open. The next instant, something went "SLAM!" on top of my head box.

"This is it!" I thought.

Something cascaded down my arms and landed around my butt. Copies of *Newsweek*, as it turned out. I heard a weird sound.

"Sluuuuuuuurp!"

Something crashed on my feet. I felt a cold, watery sensation on my ankles. The lid went BOOM! over my head, and the dumpster went black except for the shaft of light between the lids.

I didn't move. I stayed absolutely still as I heard the saleslady's footsteps fade away. After about a minute, I carefully lifted the box off my head and figured out what had happened.

The saleslady hadn't seen me. She had casually dumped a whole box full of magazines on my head, finished her cola and tossed it on my feet, ice and all.

My mouth was so dry that my teeth were sticking to my lips. I fished some cola flavored ice out of the cup and sucked it gratefully. Then I peered out of the dumpster, saw nobody nearby, and tossed out my box of loot. I clambered out, picked up my goodies, and walked away casually. A hundred yards away, I saw the saleslady discard more trash.

So here I was with a large box of books, nine miles from home. Briefly I considered securing the box to my bike and pedaling home. As I thought about this, the wind blew wastepaper and dust around my feet.

The wind had been at my back on the way to town, but I dreaded pedaling against it with a big box tied to the bike. That was when I started thinking about stashing the goodies somewhere. Being an active twelve-year-old, I already knew the location of a variety of culverts, vacant lots, abandoned buildings and so forth.

Again, I started thinking in paranoid terms. These books were so precious to me that I immediately considered an elaborate hiding place. I knew the location of an abandoned building nearby. Rather than riding my bike, I walked.

I was disappointed when I arrived at the "abandoned" building. Workmen were fixing the roof, joking around on their lunch hour. I walked right past, careful not to look wistfully at the building.

I had to use a cement culvert. A small amount of stagnant water was in the culvert, so I found a broken plastic milk crate to set the box up on. I was careful that nobody saw me, remaining in the culvert for several minutes until I could spot no passing cars on the road above. Before leaving, I grabbed several of my favorite books and stuffed them in my jacket pockets. It was hard to control myself once I started doing that, and I briefly reconsidered bringing the box home on my bike. A quick glance at the sun told me I couldn't make it home before dark if I did that. With a heavy sigh I took only half a dozen science fiction novels, including the *Star Trek* book I had first spotted. Then I walked to my bike and began the journey home.

As it turned out, my parents were heading to town anyway. We picked up the books in the culvert that very night. I could barely contain my excitement over my haul. My parents told me the whole box was, in fact, my personal property. I could sell, barter or give the stuff away as I pleased.

Little Jed, nine-years-old, listened with wide eyes. Bekka, age eleven, started to butter me up, saying, "Wow! You really did good!" She wanted a share of the books. Jed was slow to figure out Bekka's game, but once he did he made up for lost time. I was feeling generous, and told them we could go through the books and pick things out.

My dad pulled up to the bookstore dumpster when I explained other stuff had been discarded. Dad turned off the headlights as we entered the deserted parking lot.

As it turned out, a *lot* of stuff had been discarded since I had been sitting in that dumpster with a box over my head. My dad told me quietly that I was lucky nobody broke my neck. We pulled half a dozen sealed boxes out of that trash bin, some so heavy we could barely lift them.

Dad split one box open.

"Holy mackerel!" he said, looking at dozens of tightly packed paperbacks.

We pulled out all the boxes, then Jed and I collapsed in the truckbed, exhausted. Dad scooped up a coverless copy of *Newsweek*.

"I hope there's some romance novels!" Bekka said, as we roared off.

There were. It was a huge bonanza, and there were many, many such nights after that one. We kept the books we wanted, traded a few to neighbors, and bartered the rest to a flea market dealer. My little box of books was dwarfed by the big haul, but the whole night felt like it was my personal victory. I had a feeling of mastery that wasn't matched until I poached a deer later that fall. The bookstore discovery, and the many beneficial barters that followed, made me feel like a provider to my family, a male who could stand in for my father if necessary. Without inflating my ego to dangerous levels, my parents hinted as much. Of course, Dad about fell out of his chair laughing when I told him about the box over my head and the cup of cola hitting my feet.

Using A Cache

Later, I explained to my father why I had chosen the culvert rather than an old building. Dad explained to me how the culvert was actually a better choice. Poking around an old building in the dead of night could attract unwanted attention. He said the plastic milk crate was a good idea, and pointed out that elevating the box would be a good idea even if the bottom of the culvert was dry. If it rained and water ran through the culvert, my loot would still be reasonably dry — as long as it wasn't enough water to sweep it away.

When you're on foot, a cache is a great way to save effort. Rather than returning home with your find right away, or dragging your find along with you, simply hide it while you continue your route. I still prefer culverts if the culvert is large, relatively dry and available. I like to keep plastic milk crates or an old shopping cart in my favorite culvert rather than search for something to elevate my goodies at the last minute.

Remember, you don't need anything elaborate. You're only going to use the cache for a few hours or, at most, half a day.

Hedges and small, brushy trees next to walls and fences are good temporary hiding spots. Be careful not to damage shrubbery or you'll piss somebody off and cause yourself a problem. Pick an obscure, neglected spot. AND BE CAREFUL. The act of stashing something in a hiding place causes dramatically more attention and comment than rummaging in a dumpster. Don't let ANYONE see you using a cache.

If you find food that animals might damage (plastic packs of half frozen turkey wieners, for example), stash the food in the lower branches of a pine tree if available. Remember, dogs love leather, too. Using the lower branches of a tree will keep dogs away, but probably not a highly motivated cat. Pigeons and other birds will find uncovered bakery items and peck at them, but most animals *don't* find things right away. But if you stash that stuff in front of a stray dog, he'll *be* there, whether he can actually get into the things or not. Somebody will happen along and investigate what he's doing.

So be careful. As a precaution, use double boxes if available. I've used packing tape, but I don't recommend you run out and buy the stuff. Just use the stuff you find thrown out when people are moving.

Vacant lots with high weeds or mounds of refuse make good temporary caches. Just leave the stuff away from walking paths and pick it up later.

Once I left a box of bakery items on a lot of this type. When I came back, something was inside the box, moving around.

"Damn!" I thought. "Some dog found it."

That was when I heard a soft, "Quack... quack..."

I realized immediately that it was a far ranging duck from the local park. These were ducks that lived mostly on bread from park visitors. They could fly, but preferred to run or stay close to the ground. I often contemplated catching a few, but the park was always full of visitors, the fines for this sort of thing are huge — it wasn't worth the risk. Until right then.

I walked quietly to the box, blocking the opening with my body. I reached inside, grabbed, and quickly found the neck of the duck in question. As I snapped its neck, something flew in my face with a noisy flapping of wings. There were, as it turned out, *two* ducks. I figured, better the bird in my hand.

It was a nice, plump mallard duck. And, luckily, the birds had only ruined one loaf of bread, I left the bread right there and used the plastic bread bag to wrap the duck's body. That night, we had roast duck with "dumpster bread" stuffing. Come to think of it, the orange garnishes came from the dumpster, too.

If I had lived near that vacant lot, I certainly would have been grabbing ducks left and right. As it was, I had a lucky accident. The ducks had followed a trail of bread crumbs right to the box. If I had tried to do it, probably it wouldn't have worked.

Be careful about your habits, or some little kid or intrepid vagrant will rip you off. On the other hand, I've left stuff practically in plain sight and found it untouched. It's amazing. Use common sense, and with experience you'll learn the quickest, most effective methods.

In a limited number of circumstances, I've actually left the stuff in the dumpster and picked it up later. You might do this if you notice a big piece of furniture and you're on foot. Leave the loot, get your vehicle or buddy, THEN lift the thing out. You don't want it sitting outside the dumpster for several hours, attracting the attention of sharp-eyed dorm dwellers, other dumpster divers and the guy who discarded the thing in the first place.

If the item is small, however, I'd rather use a cache twenty yards away and practically in plain sight than leave the item in the dumpster. If the dumpster is against a wall I'll put the item behind the dumpster. I'm not worried about competition, I just don't want somebody to dump trash on my new finds.

Cops

That about covers caches. Let's talk about cops.

Cops piss me off. They come at you with an attitude that you are guilty and they are going to get you to admit it with a few verbal tricks. *Just once* I'd like to meet a pig that comes at me with an attitude like I have a shining aura of civil rights around my body and possessions. Criminals with guns and badges, that's all they are.

But I digress. Prior to this I have pointed out ways to avoid problems with the local swine patrol. DON'T dress like a damned burglar. DON'T carry a straight edged razor, samurai sword, meat cleaver or any other exotic blade. DON'T use matches or lighters around dumpsters. DON'T make a mess. DON'T wear clothing such as "Kill All the Pigs" t-shirts, military garb or pro-skateboarding stuff. DON'T let people see you using a cache. DON'T confront residents, employees, or otherwise act outrageous. And, of course, DON'T rummage right in front of the cops.

If you are rummaging and a cop, mid-level asshole manager, or other authority figure happens along, immediately drop your bag blade and dive stick, DON'T just toss it in the middle of the dumpster, unless you see a complex pile of waste right in that spot. In a very full dumpster, you may have the opportunity to jam your bag blade and/or dive stick into a pile of boxes. Otherwise, *hold it flush with the nearest wall of the dumpster and drop it straight down.* When Herr Copmeister shines his light in the middle of das

dumpster in his straight ahead, snout-to-the-front manner, he may miss the dropped item. Also, bags flush against the wall of the dumpster can enfold and partially conceal the item. (This is a good reason to paint your dive stick black or wrap it in electrical tape.)

Believe me, it's hard to spot a specific item in a messy dumpster unless you've been diving for a while or you know *exactly* what you're seeking. Slash told me he has hidden his diving efforts three times in this manner.

If, however, you're *positive* the police officer saw your equipment, don't try to conceal it. You'll piss that cop off, and he's probably dull and short tempered. You're cutting into his doughnut break. An attitude of mild chagrin and smiling cooperation is better than being even mildly confrontational. You can always talk shit about the police later, or seek to actually even the score. But, when confronted, the point is to get away cleanly and avoid a hassle. Don't be a slave to false pride and provoke that pig.

If you're *certain* he saw your bag blade, DON'T drop it but DON'T just stand there with the knife in your hand. Snap the blade shut.

Climb out of the dumpster if you're in it, but DON'T walk up to the policeman. DON'T make sudden moves, DON'T start offering excuses. Let the pig talk and figure out his major problem. THINK your way out, don't TALK your way out. DON'T become loud or emotional. Remain, at all times, calm and nonconfrontational. REMEMBER, you've done nothing wrong.

Tell him you're looking for boxes. This is why I like to keep several empty cardboard boxes in the back of my truck. Don't let them become rain soaked, however, or they won't make a good alibi. It's best to be familiar with the local laws regarding dumpster diving BEFORE you find yourself in this situation. More discussion on laws later.

Whatever you do, DON'T try to blather your way out of things. Cops will nod sympathetically and ply you for information. Sentences like, "I've been doing this for years and never had a problem" will produce a smile, lots of head nodding, then arrest. Minimize what you're doing. You needed a few boxes and happened to spot this nice little vase in your hot little hand. The "needed boxes" excuse is the only excuse I've known to produce a positive effect. After all, cops don't make much money, either. They're not smart enough.

By the way, always drop your loot, even if the rest of your truck is *full* of items. If the policeman says, "Where did you get this stuff?" DON'T say "from a friend." The policeman will immediately ask your friend's name and address. Remember, if the cop had enough evidence to arrest you and make it stick, he'd be pressing your face into the hood of his car. Open your mouth and say too much and he *will* have enough evidence to arrest you.

REMEMBER: *No innocent person ever avoided arrest by answering all the questions of a suspicious, paranoid pig.* DON'T fall into the trap of trying to talk your way out of things. *It won't work.*

As I stated, be cooperative and friendly, but firm. DON'T submit to search, seizure and questioning. If the cop says, "Mind if I have a look in your truck?" and you say, "No, I have nothing to hide," then you have just given that suspicious, paranoid Nazi pig the opportunity to look through your shit and draw his own conclusions. Empty beer cans? If he can shake out a *drop* you're looking at "open bottle, driving under the influence." What's that? A discarded steak knife? Looks like a concealed weapon to me. Clothes, eh? *Whose* clothes? Oh, you've got somebody's dress in your car and you don't know *whose?*

Even if that cop *knows* in his swiney little heart that this stuff is innocent discarded all-American trash, he can still satisfy his probable cause criterion, haul you in, and cause you problems. DON'T consent. Project the image of a cooperative, friendly, law abiding citizen who happens to be conscious of his civil liberties and willing to be firm, even though he is innocent and could

answer the questions if he wished. Even if you have a pile of dumpster dived goodies at your feet and the cop saw you rummaging in the dumpster, don't admit the stuff is yours. Smile broadly, shrug a lot.

Use the phrase "civil liberties," not "my rights." Cops believe that only criminals know their rights. On the other hand, they've all heard of time-consuming, cop-defeating battles over "civil liberties." So use the phrase "civil liberties," but not a lot.

DON'T say something like, "Why don't you go catch a bank robber?" This sounds a lot like a confession of a minor crime. Cops hear this a lot and it pushes their buttons, since they CAN'T catch the major criminals and must, therefore, concentrate on speeders and various kinds of thought crime. Don't be abusive, or even mildly sarcastic. NEVER call the cop names or bring up numerous instances of local piggy foul ups. If all else fails, that cop can slap you with a "disorderly conduct" charge just for waving your arms or snapping your gum too loud. Instead, play on his vanity. Say things like, "I know you have a tough, dangerous job and I hate to waste your time," or, "I'm sure glad to have your patrols in my neighborhood, officer!" If you're a female, and the cop is male, don't worry — you won't have a problem if you stroke his ego a bit. Cops are pervert sex pigs.

The only time I would use the excuse "looking for boxes" would be in a town with NO LAWS WHATSOEVER regarding dumpster diving, or in my own neighborhood. And I would only use that excuse on a really friendly cop, and ONLY if I didn't have a truck full of loot.

I wrote the following and field tested it twice on real, live porkers. It worked both times. Try it:

"Officer, sir, I respect your profession and wouldn't want you to think I'd break any laws. But I'm conscious of my civil liberties and my freedom as an American, and I don't wish to answer your questions."

Smile sincerely when you say this. Set aside all negative energy and thoughts. Don't say "sir" with sarcasm, don't act prideful or obnoxious, don't be loud. Speak to that poor, deluded pig like he's your dear, lost brother who has never seen the light of liberty. After you've said this, WALK AWAY. Don't say anything else, even if he is speaking to you, don't even turn around unless he says, "You're under arrest'" or, "Stop or I'll shoot!" or some other official announcement of forthcoming civil rights violations.

In fact, whenever possible, walk away from dumpster confrontation. Don't use that snappy one-liner, don't say *anything*, act deaf and just walk away. Wait a few days and then go back to rummaging the same spot.

Remember, from an outsider's perspective, dumpster diving is a harmless, pathetic, and not-very-profitable activity. You are hardly worth the time of a cop. You may be hassled and warned, verbally abused by pompous pigs, but most of the time you WON'T face arrest. Confrontations with cops are rare.

Good luck. Watch your rear and be prepared, just in case. Carry a quarter for that "one phone call."

More Etiquette

Before we conclude this chapter, I'd like to share a few more tricks with you about dealing with other dumpster divers.

I figure a dumpster is rather like a store shelf. Just because somebody parks their cart in front of a sales special doesn't mean it's all *theirs*.

If somebody is rummaging in a dumpster, DON'T intrude on their "intimate personal space." DON'T touch them. But feel free to reach in and grab. If the other person says, "Hey! This is mine!" just smile broadly and keep rummaging. DON'T wage a war of words, even if they start telling you about the rock hard nature of their luck. If you have a vehicle, toss the crap on your truck. Otherwise, drop it at your feet and stand over it. Putting stuff in a box, if available,

establishes a more definite claim. NEVER try to grab from somebody else's hands or their "pile." Eighty percent of assaults in a psychiatric setting occur when somebody tries to take an object away from the patient in question. Most vagrants are psychiatric patients, so DON'T grab stuff out of their hands or their pile. You're doing this to profit, but the other guy may be doing this to stay alive. Don't assume he is stable. Don't underestimate his level of motivation.

If the other dumpster diver is on the right side of the dumpster, start at the left and work toward the middle, dominating as much space as you reasonably can. Climbing inside the dumpster will give you more opportunities to grab and more chance to dominate the dumpster space. If you have a partner, both of you should rummage. You can pick up the stuff and toss it on your vehicle AFTER it's all in your pile.

If you are the one rummaging and somebody else approaches, use these tactics in reverse. Move to the middle of the dumpster or climb inside. If they approach, say, "You *mind?* I was here *first!* Go somewhere else, I got *this* spot!" Say this in an assertive, firm manner but not as though you are willing to become violent. This is the one exception to the rule NO VERBAL CONFRONTATION. Why? Because it works. You were there first, you're dominating the dumpster, most people will back off. Tossing in a remark like, "I got five kids to feed!" works well, too, but only use this as a second line of verbal defense. It gives them an opening to justify their intrusion. DON'T try this tactic if you are outnumbered and/or the people look big and strong enough to cause you a problem. This works well on most vagrants, who are a timid lot, though most vagrants wouldn't approach in the first place.

If you have a partner, he should stand in front of the dumpster in a protective manner, facing off the competition, while you toss stuff out of the dumpster, either into your vehicle or behind your partner.

DON'T get into a battle of words. Repeat once if necessary your position: A) Here first. Go elsewhere. B) Five kids, the youngest needing surgery.

Say this firmly, not TOO loudly, but with true sincerity. YOU should believe it. I've seen Slash almost weep as he named all five younger siblings and described the needed surgery in gory detail. But that's Slash. He has a knack for bull-shit.

Remember, YOU may be a self-reliant individual who doesn't waste pity on folks who can't cope with the cold, hard universe. However, the "five kids" approach works well on people who receive welfare because they *need* it and never heard of Ayn Rand.

Let the competition talk, threaten, beg, show you kiddie photos, doctor's orders, etc. Remain silent and firm. If you can't stay silent, say, "No!" Say it a *lot*. After you've *cleaned that dumpster out*, toss them a few "Hungry Man" dinners or something. Say, "No hard feelings, buddy, but I got kids to feed." This is not a concession to pity. I believe that if you can avoid having a mortal enemy for the price of a few discarded pot pies, pay that small price. But DON'T negotiate a split. Dominate those dumpsters and throw the competition crumbs, not concessions. Only the lean and hungry shall survive.

After a bad confrontation, lay low for a few hours. And, if you can, get rid of all your loot before you dive another dumpster. The competition might be calling the police, claiming you and your vehicle have committed various crimes. DON'T be tempted to do likewise. Calling the police in order to get somebody else in hot water seldom works and often backfires. NEVER call the police and claim your least-favorite dumpster diver is making a big mess. You'll just start an anti-dumpster diving program that will hurt *you*.

Some dumpster competition is no big deal. I get pretty hard assed about FOOD, but I've had some nice conversations with people while salvaging books from library dumpsters. I also had some nice conversations with an old lady while diving a wholesale florist dumpster. She did,

however, try to lay a guilt trip on me, saying she was giving flowers to her friends in the hospital while I was just giving the flowers to women I wanted to date. I saw her at a flea market some weeks later, selling pressed flowers, potpourri, used pots and small, carefully pruned vines.

So play it by ear where competition is concerned.

If you see graffiti inside dumpsters, don't be intimidated. Ignore it or slap a SUPPORT LOCAL POLICE bumper sticker over the graffiti. In a residential area, the locals will think some other resident did this. NEVER put graffiti on a dumpster to mark it as "yours." This doesn't work and only causes hassles.

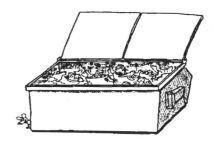

TWISTED IMAGE by Ace Backwords ©1993

WELCOME TO "SISKULL AND EGGBERT AT THE DUMPSTERS"

TODAY WE'LL BE REVIEWING ONE OF THE HOTTEST DUMPSTERS IN THE NATION!!

WELL, I FOUND DUMPSTER DIVING TO BE A POWERFULLY EVOCATIVE EXPERIENCE!! NOT ONLY THAT, I FOUND ALL SORTS OF NEATO JUNK!!

I DISAGREE, ROGER... FRANKLY, I COULDN'T GET OVER MY DEEP-ROOTED FEELINGS OF SHAME AND FAILURE THAT I ASSOCIATE WITH DUMPSTERS! PLUS, I SMELLED POO POO...

WELL, I THINK THIS DUMPSTER BELONGS IN THE DUMPSTER!! THUMBS DOWN!!

I LAUGHED!! I CRIED!! I WALLOWED IN FILTH!! THUMBS UP!!

Chapter 7
The "Big Three" Dumpster Hot Spots

Remember the "maximum profit, minimum effort" approach: grab the good stuff first! In keeping with that spirit, allow me to present my "big three" dumpster hot spots, followed in the next chapter by my "lucky seven diving sites." If you dive *only* the "top ten," you could still have a "maximum diving lifestyle."

#1 With A Bismark — Bakeries!

For sheer dependability, combined with value and volume, bakeries can't be beat. Each day bakeries produce thousands of doughnuts, loaves of bread, pies, cinnamon rolls, cookies, cakes and specialty items. If the stuff doesn't sell the first day, or in a couple days at the most, *out it goes.* The problem isn't finding the stuff, but consuming it. One bakery can supply a small commune.

Like all dumpster loot, the stuff you find isn't necessarily a cross section of what the business sells. Seldom will you find those tempting, crispy French cookie things covered with dark chocolate. You know, the kind you'll gladly purchase for 60¢ apiece, just for a *taste.*

Bakeries will provide more "two day old" bread than you can handle.

Here's what you *will* find: BREAD. Enough to supply you and yours with toast, buns, sandwich fixings, stuffing, croutons, French toast, bread pudding and much, much more. Enough to feed a handful of hungry swine. Rye, pumpernickel, wheat, white, French, Italian, raisin and sourdough, TONS of it over the course of a year.

Sometimes you'll find lots of one kind, then you won't see more for months. Often you'll find the same thing day after day. Once, we kept finding so much cheese bread that I was tempted to walk inside and say, "What is the *problem*, folks?" (Don't do that, of course.) We ate the stuff, we bartered it, we fed it to hogs and chickens and still the damned cheese bread was gaining on us. Slash would look at me tiredly as we loaded up our truck and say, "Brother... there's something cheesy going on here."

We finally had to run out and acquire more chicken and swine just to keep ahead. But what a pleasant problem to overcome... *too much* food.

In most cases the bread isn't "stale," merely less than oven fresh. For reviving stale bakery items, either pop in a microwave or sprinkle lightly with water and warm in the oven. But even less-than-fresh bread has value. In fact, stale bread is better for French toast.

Expect also to find caramel rolls, doughnuts, cookies, and pies. In my experience you'll find few cherry pies, lots of apple, raisin, chocolate and pumpkin. Does the bakery make "gross looking" chocolate pies? Expect plenty. You'll find things that actually taste better than they look or sound. I wouldn't rush into a bakery and buy "chocolate pineapple chunkies." However, when I find something like that and sample it *ala dumpster*, I find it tastes *good*. I'm tempted to walk in the bakery and say, "Hey — why don't you call these things 'tropical chocolate treat-a-roos?' You won't end up throwing so many in the dumpster." (I never do that, of course.)

Expect more pumpkin pies around Christmas and Thanksgiving, as well as more cookies shaped like turkeys, bells, Santa, etc. Cakes are more rare in my experience. I suspect this is because many bakeries make cakes on a pre-order basis and leave the stuff on display longer. When I do find a cake it's often in bad shape,

good only for a few quick bites and the rest is a treat for the hogs.

Sometimes you'll find bakery "screw ups." Once we found some chocolate cookies which were so fresh the whole cab of the truck was filled with a delightful aroma. Slash and I couldn't resist — we pulled over and started stuffing our faces.

We only had a mouthful apiece when we looked at each other, dismayed.

"Bleccch!" I said, spitting the stuff out the window.

Slash didn't have the window down on his side. He spat right on the floor.

"Whuh da fuuuck?" he asked, scraping his tongue.

"No sugar in the damned mix!" I said, looking frantically through our "blizzard kit" for our emergency fruit juice.

"We ought to go back and complain!" Slash exclaimed.

(We didn't, of course.)

We carted the stuff home for the hogs. However, only a few weeks earlier we had found dozens of cans full of ready-made frosting. Dad smeared the stuff on the sugarless cookies and pronounced the result quite good. Slash and I passed, thanks.

Stale cookies are just fine when dunked in milk, or ground up for pie crust or ice cream topping. You'll never manage to use all of them, however, unless you have eight kids or a commune.

Often you'll find wasted raw materials. If you gather up ten or so 50 lbs. flour sacks, and shake and pound them over a large container, you can often salvage a pound or so of flour. Sometimes you'll find ten or fifteen pounds of flour, wet at the bottom of the sack or "contaminated" by chocolate syrup. Often you'll find plastic bags with twenty pounds of "bad" dough, pouches full of fruity filling, poorly scraped gallon containers of chocolate syrup, and so forth.

Slash always thought it was amusing to pick up a half-used cone of whipped cream, place it between his thighs, and say, "Look!" When his unfortunate partner would look, Slash compressed his legs and squeezed the cone with his hand, producing an obscene blast of whipped cream. I hated it when he wasted stuff like that. I always liked to "shotgun" it into my own mouth.

When you find a large amount of dough, take it home and let it rise for an afternoon. Keep it covered with a wet piece of linen. Bake the whole thing in one humungous loaf and feed it to hogs or chickens. (*Don't* feed animals raw dough.) In hard times, bake and eat it yourself in suitable portions.

Don't eat stuff directly from the dumpster. It's hard to resist, especially at bakeries, but always check things out carefully first and, again, don't attract attention and comment.

Bakeries are more predictable than most businesses, and discard their goodies in a routine manner. I find this is often done at night, shortly after the close of business, or first thing in the morning. Hang around the bakery until closing time and see if you notice the workers grabbing large amounts of stuff from the "day old" and "bargain" shelves. Stuff from the other *shelves* is probably making its way to the back to be packaged as "bargains." You can hang out across the street and watch for the same thing. If you can find out the "discard time" you can usually count on it within an hour or so. Remember, the days around a holiday are *hot*, especially when the bakery is closed and can't sell off its day old stuff.

But WATCH OUT! Bakeries can be in full swing at three o'clock in the morning. Be quiet. Remember, also, that many stores have "in house" bakeries. These places can be even more lucrative than a regular bakery, because they are one component of a business and wasteful prac-

tices can be absorbed by the rest of the operation. The "cheese bread" dumpster was behind an in-house bakery operation. Doughnut shops are also an excellent and dependable dive spot. The bigger, the better. Such places are more profligate in their waste. But I've had excellent luck with small bakeries, too.

Don't delay. Get yourself "a piece of the pie."

#2 Grocery Stores, Supermarkets, Food Marts

Happy divers display a moderate haul of grocery store goodies.

You may never shop the same way again!

Given a choice of diving only one kind of dumpster, I would choose grocery stores. Bakeries can't be beat for value, volume and dependability. But for value, volume and diversity, grocery stores are in a class by themselves. Sometimes you'll find nothing of value. Other times you'll make out like a bandit. Because of the tremendous variety in this particular dumpster, I've used categories in this entry.

Discarded Produce

Produce found behind grocery stores can require quite a bit of sorting.

You can count on grocery stores to provide you with fruits and vegetables on a regular basis. The number one item, in my experience, is GRAPES. After that, count on bananas, apples, pears, peaches, and grapefruit. Frequently, these delicious bonanzas "hide" beneath piles of plastic, unconsumable produce refuse, outer lettuce leaves and so forth. If you dig into something and notice one or two respectable pieces of fruit, just grab the whole box. A great deal of sorting and trimming is required to make the most of discarded produce, but it's easy and worth it.

You can expect almost anything. My favorite vegetable is the artichoke, but I never actually purchased one until age 26. Expect seasonal cycles, of course. When the store is full of watermelon, so are the dumpsters.

Citrus Fruit. Citrus fruits often take a beating and end up in the trash bins. This happens to a lot of fruit, but it seems to me that citrus is more vulnerable.

Oranges, in particular, are wimpy fruits. I often find whole crates covered with greenish

mold, useless. Now, mind you, a little mold never killed anyone. If you see a tiny piece of mold, rust, mildew, rot or bruising on a piece of fruit, just trim it. The American obsession with pure produce drives me nuts, honest to God, even though I thrive from it. Seriously, you think those people on wagon trains chucked their food supplies whenever they found a green speck? Of course not. Mold was around for millions of years before *homo sapiens*, and as we evolved we have been eating the stuff. I believe that decades from now people will take "mold and mildew supplements." Of course, that's not what the people marketing the stuff will call it.

Anyway, I've taken five pound blocks of cheese completely covered with furry colonies of mold, trimmed it into a respectable three pound block, made myself half a dozen grilled cheese sandwiches AND I'VE BEEN NONE THE WORSE. Every day you consume small amounts of mold without becoming ill or dropping dead. It's ironic that people will shun something basic and natural like mold but will gladly eat foods laced with chemicals. You can't live your life in a germ-free bubble. People never exposed to anything have wimpy immune systems. People exercise their bodies and minds, but — except for receiving a few shots — few people make an effort to exercise their immune system. They're afraid they will "wear themselves down." Like an army, an immune system requires drills to become proficient.

But I digress. Often I find oranges that look like tennis balls. These are good only for compost. But lemons, limes and grapefruit seem to endure better. The fruit may not be fresh enough for table display, but it is still great for juice. I'm a firm believer in the value of a juice extractor — but don't pay full price.

Manage your haul carefully and you should always have a stockpile of citrus fruit and juices. After a hard morning of chores on the farm, the Slashmeister and I would take a break and slam down a quart of juice apiece. This helped a lot when we boarded the bus with all those sick children whose parents wouldn't dream of poking about in an "unsanitary" dumpster.

Non-Citrus Fruit. Grapes, as I stated, lead the list. When the bunches arrive, employees trim them into small bunches — the kind you buy in the produce section. They discard many small strands of grapes — anywhere from six to three dozen. These "reject" bunches are discarded by the crate. Some are small, some discolored, some have broken skins. Don't fret. Run the grapes under cool water (do this with all fruit) and pick off the worst-looking grapes. Don't worry about a little bit of brown on the ends. If your apple begins to turn brown while you eat it you probably will still finish the apple. NOTHING IS WRONG WITH THOSE GRAPES.

Eat and enjoy. Very small grapes are often sour or extra-sweet. One tiny grape may be sour, and the next one on the bunch sweet. I like to watch CNN with my teeny tiny dumpster grapes, breaking the skins carefully with my teeth. When Jed was really little, and a bad taste in his mouth was enough to make him bawl, Bekka and I would triple dare him to sample teeny tiny grapes. He would eat happily for a while, then start to wail. Jed recalls those days fondly.

Homemade raisins are good, too, especially when you find too many grapes to eat fresh.

During the Chilean grape scare some years back, my parents brought home *four truckloads* of grapes. Some of the stuff wasn't Chilean, but grapes that went bad because people wouldn't buy grapes. Realizing the stuff was being thrown out all over the country, my parents went to several stores we rarely dived and even another town we seldom visited. The solution which sprang to mind, naturally, was to make juice. We also made several bottles of wine, just for fun. The thought that the stuff might actually be poisoned occurred to us after several hours of sorting. We solved that problem by feeding the grape pulp to the hogs after we juiced it. The hogs didn't keel over and neither did we.

Most food scares are bullshit. Whenever we hear some crap on t.v. about "contaminated" or "poisoned" food, we fire up the truck. Once, an employee asked Slash if he wasn't afraid the stuff we were loading on our truck wouldn't make us sick. It was, as I recall, some crap about ice cream bars that may or may not cause a certain influenza.

"Hey, buddy," Slash said. "Watch!"

With that, he unwrapped an ice cream bar and shoved the whole thing in his mouth. The employee looked at Slash as though Slash had just swallowed arsenic.

"Thagh whlzh dub nob kuiw muh shtrumphemph muh!" Slash stated.

"What did he say?" asked the wide-eyed employee.

"He said," I answered, "that which does not kill me strengthens me."

With that, I casually consumed a "deadly" ice cream bar myself.

Again, however, avoid outrageous behavior even when provoked by stupidity.

Back to fruit. Peaches, pears, apples, plums and apricots can be bruised or even rotting a little and still have good, edible portions. Trim and eat, use in pies or other cooking, or dry for future use. Apples are good for apple butter and cider. DON'T try to run "mushy" apples through a juice extractor. You'll get mush. Exotic fruits like papaya and guava frequently find their way to the dumpster. Enjoy, enjoy.

Bananas are a favorite of mine. You will frequently find whole boxes of "rotten" bananas, so ripe they've turned black. These make excellent banana bread. Grab it and growl!

Watermelon is wasted in tremendous amounts every summer. You'll find lots of individual slices wrapped in plastic as well as broken or cracked whole melons. Trim until you find firm flesh. Chill and eat. Delicious! Chilled watermelon juice is refreshing, and easily produced from "less than firm" flesh. Little children like this stuff during the summer more than thick, tart juices. The seeds are good roasted and have an exotic flavor.

Other melons are wasted in delightful abundance, including honeydew, cantaloupe, "musk" and bitter melon. One great thing about a melon is size. Trim a peach and you've got three quarters of a peach. Cut the spoiled patches off a melon and you've got a *meal.* So when you find a whole box of melons, you're in business. For a delightful dumpster treat, make a fruit tray featuring grapes, watermelon and honeydew. Delicious!

Berries, especially strawberries, are wasted in huge amounts in their proper seasons. It may seem like a lot of work to trim individual strawberries, but have you *priced* the darned things? It's well worth the effort to sort and trim the darlings. Don't eat too many while you trim or you won't have enough for strawberries and cream! (We'll obtain the cream for you under DAIRY FOODS.) If you have several crates, which often happens, use in pies or trim and freeze in plastic bags. Don't buy bags — old bread bags work just fine. (Despite a B.S. "paint" scare some years ago.) Store abundance away for a rainy day. Strawberries add a kick to other juices, too. I like a strawberry garnish on a nice glass of carrot-apple juice, all components *gratis,* of course.

Raspberries, blueberries, gooseberries and other little delights are discarded more rarely. Don't let a few smashed berries discourage you. You'd eat it if you had just picked it.

Cherries are discarded extravagantly in their season. Better to pick out the worst ones and set the fruit out for consumption — let the person eating the cherries do the trimming. Don't fret about a few brown spots. Cherries also add a "kick" to other juices and are good dried.

Pineapples are discarded more rarely. When I find a discarded pineapple the outside of the fruit is often dark and/or feels "slimy." It's still good inside, however. After you eat the best part of the pineapple toss the tough "core" in the juicer. It's good.

Bruised figs are excellent, even the bruised parts. I rarely find dates, however. This about covers most of the major fruits. There are plenty of other things you'll find, such as pomegranates and avocados, kiwi and star fruits. Simply trim the "bad parts" and use. With large amounts, keep an eye toward preservation.

Grocery stores discard lots of lettuce and cabbage leaves suitable for animal feed.

Dive Your Veggies! The vegetable you'll find most frequently isn't fit for human consumption. That vegetable is lettuce. No, not succulent heads of lettuce but the outer leaves. These are routinely discarded by the boxful. Hogs will eat the stuff if nothing else is available. Chickens will peck at it half-heartedly. But rabbits and goats love it. We raised a hundred or so rabbits and they required little more than these discarded leaves, water and a salt lick.

As an experiment, we once made homemade "kimchee" out of those tough discarded cabbage leaves. It was excellent. The succulent inner cabbage leaves turned out too soft. So it's nice to know that when economic collapse takes place, you'll probably still be able to find discarded cabbage and lettuce leaves.

Cabbage and lettuce are sometimes discarded by the head. Peel off the wilted outer leaves and eat. By the way, no matter how wilted a vegetable might be it still tastes good boiled or steamed thoroughly. In fact, it's tough to tell fresh boiled stuff from stale boiled stuff. Remember that, because you'll find a *lot* of wilted stuff. Artichokes, for example. A buck or more apiece but you can have 'em for *free*. Don't be discouraged by brown tips on the leaves, or brown wilt on lettuce. Just don't eat that part. After all, the thing grew in the dirt. Don't be a wimp. Don't trim off the artichoke stem. Boil and eat. It's good, too.

Speaking of growing in the dirt, many excellent root vegetables are discarded simply because they are too small. You'll find carrots, radishes, beets, turnips, parsnips and other root vegetables with their tops still attached. Other than a few blemishes, nothing is wrong with these delicious vegetables except their size or freshness. Little bitty veggies smothered in butter — mmm , mmm! Need I say more?

Often you'll find "assortments" of veggies in plastic wrap, labeled "stir fry" or something. This is dumpster convenience at its finest. Take a gander at those price tags and pat your own back heartily.

You'll frequently find wilted spinach, collard greens, and other leafy green veggies. Boil or steam and enjoy. Wilted celery! Trim the brown spots, chop it up and toss it in a stew. Throw in the leaves, too. They're good.

Cucumbers aren't as tough as they look. I find lots of them, bruised and banged and even split in half. Trim carefully and cut into cucumber coins. Leave the skin on, it's the best part.

When you find tomatoes they are usually in bad shape. Don't fret. Carefully pick out the worst parts and make spaghetti sauce with what's left.

Potatoes are often found in a somewhat dried-out state, frequently with "eyes" on their outer skins. Wash 'em, trim 'em, fry 'em up skin and all. If you see the top of a potato or onion bag in a dumpster, pull it free. These vegetables are frequently discarded in their bags.

With onions, you may note black spots or soft areas. Simply trim. Treat garlic bulbs in the same manner. Naturally, these vegetables aren't very expensive. But every little bit helps. Use the money you save to purchase a big, thick steak.

Often, you'll find broccoli which has become slightly yellowed, wilted, or has a "slimy" texture. Slimy vegetables are generally the result of those stupid misting machines in the produce section. These things don't keep vegetables fresh, merely fresh *looking*. The veggies actually go bad quicker as the constant water action sucks the life out of their degenerating cells.

Well, don't fret about slimy broccoli. Simply trim, preserving as much of the tender upper stalk as possible. Then chop up the whole stalk, including the leaves. This is really good, and good for you. I actually prefer the stalk to the broccoli tips. You've probably eaten so-called "tough" stalks of broccoli in an oriental restaurant and enjoyed it.

Sometimes you'll find nothing but whole boxes of broccoli stalk. This is more common behind restaurants but I've seen it happen at stores, too. When I find nothing but the broccoli shaft I cook it up and call it "Broccoli ala Bush." Read my lips — it's delicious!

I'm a firm believer in using the whole piece of produce whenever possible. When I put a melon in my juice extractor I leave the skin on it. Slash and I eat every part of an apple but the seeds and stem, and Slash likes to eat small shreds of orange skin with his orange pulp. Many people I meet are shocked that I have such good nutritional habits and yet I do this nasty dumpster thing. On the contrary — dumpster diving has always been an aid to my nutrition. Can you imagine supplying three kids with all the fresh produce they can eat — especially if they really develop a hankering for the expensive stuff? But I've always had every kind of food I could desire.

Back to veggies. Sometimes you'll find sweet corn still in the husk. When you peel back the husk you'll notice black kernels, soft spots, etc. Don't fret. Take a knife and cut, cut, cut. The corn looks funny when you serve it, as though somebody came along and tasted each cob. When you find a lot of corn, strip off the husk and freeze it.

Around Halloween, expect pumpkins. Use in pies, bake like squash, roast the seeds. Don't forget about seed "windfalls" from produce.

Besides the veggies already mentioned, you will frequently find green beans, fresh peas, bean sprouts, cauliflower, asparagus, and all types of peppers. You will rarely find dried beans such as pinto beans, but I've seen it happen. I've also seen such exotic and expensive veggies as French endive. Now *that's* gourmet dumpster diving.

Remember, produce is frequently discarded in boxes which carry the name of some fruit or vegetable. Don't be dismayed by a "crappy" appearance. Carefully separate trash from treasure. Remember, safe inside its "natural packaging" that produce is clean and delicious.

Other Produce. Besides fruits and vegetables, there is another category which includes such things as coconuts and cashews. Perhaps I would mention it only in passing if it wasn't for *coconuts*. And stale peanuts in the shell.

I find coconuts constantly. Whenever a customer drops a nut and busts the shell, out it goes. Other customers feel compelled to poke the coconut "eyes" until they pop. Then the juice drains out while the customer trudges off to the canned pasta section. When you find a coconut smash it open, pull the meat out, shred it and toast for cookies. Don't trim the brown layer off the meat (where it touches the inner part of the shell) because it's so *good*. You can also eat coconut raw for a chewy snack.

My dad used to love raw coconut, despite the fact it took him half an hour to eat even a small slice.

"It prevents beri beri and dysentery!" he would exclaim.

So far as I know, none of the Hoffmans suffered from these diseases while we had fresh dumpster dived coconut. And a little goes a long way.

A few times we made huge hauls of stale peanuts. These can be toasted and "revitalized" or made into peanut butter. Infrequently, we have found other nuts such as cashews. But these are better dived at specialty shops.

Be aware of odd produce. Once Slash and I found a box full of "petrified" parsnips. We didn't even bother to grab 'em for compost. They smelled terrible. Later, we found out what we had foolishly left in the dumpster — horseradish root. We felt sick at our mistake. Mom loves horseradish with meat and the opportunity to trade it with neighbors would have been profitable.

Another time we found some cactus leaves. We figured somebody in the "potted plant" section had trimmed these leaves off some ornamental cactus. A few weeks later we found out these were tender cactus leaves intended for human consumption. I didn't have the opportunity to sample this delicacy until years later when I moved to Texas.

From these mistakes I learned to take an interest in anything new or different in the produce section. When a Chinese student moved from his apartment and tossed out most of his pantry, I quickly recognized dried straw mushrooms, seaweed, black fungus, and a variety of other odd-but-edible items. Take an interest in food and you will do well. What could be more fascinating than your next meal and where it's coming from?

"Expired" dairy products are a frequently discarded food item. This dumpster divin' dog knows he'll soon get a treat.

Dairy Products — The Sour Smell Of Success

Thank goodness for power outages, refrigerators with bad motors, and other things that curdle the cream in the night. Thank heavens for expiration dates — and mold. Some of my best friends are spores.

Dairy products such as milk (skim, homogenized, 2%, chocolate, etc.), cream (including whipping cream and half 'n' half), yogurt, and all that lovely cheese and pseudo-cheese have a wonderful habit of going bad or remaining in stock past their so-called expiration dates. Expiration dates are mysterious dates when perfectly good food mysteriously transforms into deadly poison at the stroke of midnight. Yeah, right. And my pet rabbit does book reviews. Of course, when I find expired food still in stock I complain vigorously to the management. I've also been known to file complaints with my local health department.

Often, "expired" milk isn't even sour when discarded. However, if the weather is warm it soon will be. If you see discarded ice or dry ice from the meat department, pack it around your salvaged perishables. Cold weather is excellent when you find meat, frozen foods and dairy products. It thins out the riff-raff, too.

Usually you won't find one or two gallons of milk — it's more like ten or twenty gallons at a shot. Just because milk is sour doesn't mean it's "bad." Use for cooking. Refrigerate it as soon as possible and use it while it's fresh. Encourage your kids to drink a gallon a day when it's available. Feed it to cats and dogs. They won't bitch because the milk is "sour." Animals are smart that way.

Trim moldy cheese and consume it with confidence. Novices may feel more comfortable making grilled cheese sandwiches, but there's no reason you can't trim and eat immediately. Invisible mold spores will quickly recolonize the cheese if you leave it sitting around too long after trimming. Don't be discouraged by cheese that looks like a furry brick. As long as some cheese is still visible the stuff is probably good — especially if the brick weighs several pounds. When you price a five pound slab of cheddar, Colby or Swiss you'll pat yourself on the back.

Cottage cheese holds up amazingly well when exposed to summer temperatures. Drain off excess whey before consuming. If you find frozen cottage cheese in the winter, the stuff will "break down" when thawed. Don't worry about it. Use in place of ricotta cheese in lasagna.

Of course, when I think of dairy foods I think of ICE CREAM. Lots of this wonderful stuff is discarded, everything from plastic gallon buckets of butter brickle to chocolate-covered ice cream bars to boxes of Neapolitan. Despite my anecdote, food scares are rarely the reason this stuff is discarded. It is tossed out as commonly as milk. However, it's a little trickier to use.

If you obtain the stuff in a frozen state toss it in the freezer and pat yourself on the back. But if you obtain the stuff in a severely melted or a liquid state, more work is required. As you may have discovered, ice cream does not return to its previous state when melted and refrozen. So you'll enjoy the stuff more in shakes. Use a blender. You can also convert severely melted ice cream bars into chocolate shakes in this manner. Be careful to remove all the sticks, of course.

Some ice cream bars and other "frozen treats on a stick" can be refrozen and then chipped out of the box for individual consumption. This works especially well with snacks that are all one "component" like fudge bars. It also works well for stuff with a chocolate "sheath" which holds the thing together, even in a melted state. However, a lot of frozen treats separate when melted and refrozen. You might pull out a banana fudge pop, refrozen for consumption, and find a thin, icy "banana" flavored section on one side, and the other side is a sludgey, frozen "fudge" flavor. Yech. But you won't know until you try. Some refreeze successfully.

Melted ice cream in a plastic or paper bucket is easy to deal with, but the stuff in a box is harder. If you pick it up, try to open it, etc., it can turn to glop and run through your hands. Best to freeze it immediately, handling it as little as possible. If you have a huge freezer, toss the whole box inside. Don't sort, don't separate, don't poke around too much, just freeze the whole thing. When it is frozen you can peel off the packaging and utilize it.

Once, Bekka and I returned home with four large crates of melted ice cream. I would suppose each crate contained sixteen boxes of ice cream — liquefied.

"What should we do, Mom?" we asked, showing her the boxes.

Mom scratched her head. thought a moment, then said, "Were going to put off butchering that hog a few days."

I thought her reply meant we should feed the stuff to the hogs, but then she directed us inside the porch with the first of the crates. She lifted the lid of our huge storage freezer, and I was surprised to see it was almost empty. Usually it was nearly full.

Of course! I thought. She moved everything around so we could put the hog meat in the freezer. But, instead, we tossed those four crates inside.

As we moved the crates, our farm cats scurried around lapping up the melted ice cream oozing from the boxes.

"Let's field test the stuff," I suggested, and tossed a gloppy box of vanilla toward a group of kittens.

They loved it. After they finished licking that box completely clean, inside and out, they licked each other clean the rest of the afternoon. It was quite amusing.

A few days later we "peeled" everything and put the ice cream in more convenient containers. And, of course, we made lots of shakes. Most of the ice cream was Neapolitan and it had retained its striped appearance despite being liquefied. Jed found the slightly warped stripes amazing. He kept saying, "Wow...wow..."

Dumpster diving is a whole universe of "cheap thrills."

Sour cream and various dairy-based dips are thrown out with amazing frequency. I can even tell you the two most commonly discarded types: French onion and garlic. Of course, spicy avocado is gaining rapidly on the two leading dips. Even when exposed to warm temperatures for several hours, these dips hold up amazingly well. You may find a few tablespoons of whey on top of the dip. Simply drain it. Use these dips with chips or as baked potato toppings. Of course, I like sour cream so much I can sit down and eat it with a spoon. Hope you love dip, too. The problem isn't finding the stuff, but using it.

And you'll find a lot of yogurt, too, especially pineapple. Hope you like that flavor. A lot of pineapple, lemon and lime flavored things end up in dumpsters. My dumpster diving has led me to the following insight: Americans don't like yellow things or lime green things. They *especially* don't like yellow or lime green slime textures. I think it all goes back to slapping their little hands when they pick their noses.

Frozen Foods — Dumpster Gold!

A huge variety of frozen foods are available in our society, everything from two-for-a-buck pot pies to gourmet entrees. All these delicious delights have expiration dates. And they are all wonderfully vulnerable to freezer burn, to being crushed and ripped and dropped. The dumpsters are full of this expensive, wonderful stuff.

Pot pies and frozen dinners of all sorts are among the most commonly discarded items. If you find one, you'll often find twenty to sixty. I have seen something near a truckload, *hundreds* of these delicious and filling delights. If you find six or twelve, you can toss 'em in the freezer and use individually. However, when you find a *lot* and start to run out of freezer space, simply break the meals down into individual components. For example, take the t.v. dinner and scrape the vegetables into one container, the fried chicken into another, the potato and gravy unit into yet another. Tupperware is wonderfully useful, but don't run out and buy the stuff. Plastic ice cream and margarine containers work just as well. If they become warped and won't seal properly toss them out and get more.

Regarding frozen food, remember this: the stuff has been frozen for a long, long time. It was packaged under reasonably sanitary conditions. It won't kill you just because it sits thawed for a few hours.

Various types of frozen dough are exciting items. When the stuff begins to thaw it begins to expand. Sometimes one of these "cardboard cans" will go pop! right in your hand. I have picked up a large box of frozen dough and heard up to half a dozen soft pop! sounds.

"They're playing our song!" Slash would say on these occasions.

If you find dough in a frozen state, toss it in the freezer for later use. But don't freeze it if it has thawed and started to "rise." Let it rise and bake according to the directions. Fresh baked cinnamon rolls straight from the dumpster! How's *that* for delicious convenience? Most instant cookie dough doesn't "rise," so just refrigerate when you obtain it.

A variety of frozen, breaded meats can be obtained in dumpsters, such as veal patties, fish sticks, chicken nuggets, and so forth. If these things are thawed, cook 'em extra good before consuming. These meats, however, are not discarded as frequently as the "meal" and "entree" packages. And, oddly enough, I rarely see frozen vegetables or stuff like hash browns and french fries. Perhaps these have a longer shelf life, or sell quicker because they are cheaper. When I do see these items it is usually in amounts of one or two and damaged in some way.

Keep in mind food is discarded for a variety of dumb reasons, including expiration dates, damage or convenience. If a customer rips open a box to see if the "Hungry Dude" dinner contains extra cheesy weenies, then leaves the item on the shelf, that item won't sell. Often, the *last* item on a shelf won't sell.

If a product has been discontinued, and half a dozen "Turkey Gizmos" continue to monopolize shelf space, a manager might simply say, "Chuck 'em!" That is an example of discarding for the sake of convenience.

I make an effort to determine why food has been discarded, and I'm especially careful about meat. Usually you can tell by looking at the package. Has it been opened, ripped, crushed or otherwise damaged? Next, check the expiration date. Lastly, ask yourself if you've ever heard of "Mr. Rudy's Extra Spicy Clam Chunks 'n' Brisket Bits." Maybe this is a discontinued product that sat around for too long. But don't be paranoid, don't start making phone calls or playing Sherlock Holmes. Most things are discarded because somebody in management is an idiot. Or a chicken. If I were running the store I'd say, "Eh, yeah... just load that stuff right in the back of my truck. Make double sure everything is really expired, in case the general manager asks. Oh, and take an entree home for yourself, too."

Most managers are so afraid of a "pilfering" accusation they won't take advantage of these goods. This is stupid, since I've noticed most managers actually pilfer. They'll fear for their soul over a box of paper clips but God forbid they should salvage perfectly good food.

Frozen juice concentrate is often discarded. Don't worry if it has thawed, just refreeze. Of course, things like this may leak, and the containers are often splattered with food refuse. A stain on a package is considered "damage," and most people won't buy the stuff. And here I thought packaging was to "protect" food. Go figure. Anyway, wiping something clean or removing a juice stain from your freezer is a small price to pay for juice concentrate. But don't spill the stuff on your carpet. Grape, especially, becomes a permanent fixture. So handle with care when the stuff inside is liquid.

Considering their cost, frozen foods are dumpster gold. And they are a commonly discarded item. Assume the stuff is good, because it usually is good. Price it, and pat yourself on the back.

Canned and Dry Goods — Primo Dumpster Loot!

Because of the durable nature of canned and dry goods, these items are not discarded as frequently as produce, dairy or frozen foods. However, I've seen many hauls which involved nothing but canned and dry goods. I've seen hauls of a hundred salvaged cans, a hundred boxes of cereal and macaroni. Many people assume dumpsters contain very little usable food, but I tell you that one of your biggest problems is *dealing with the surplus.*

Cans are at least as common as boxes, but I become more excited about canned foods because these items can be stored as part of an emergency stockpile for a long, long time. Also, frankly, lots of yummy things are in cans and I have fond memories of all the fun I had as a child with "mystery cans."

"Mystery cans" are cans without labels. Naturally, the only way you can determine the con-

tent is to open one or speculate intelligently. In the case of two dozen different sized and shaped cans this can be quite a puzzle.

I have asked grocery people why a label would be removed from a can before the can is discarded. One told me, "To obtain a rebate." Another said, "Maybe to keep count of what they threw away." Yet another said, "To prevent re-sale." When my brother was a stock boy, he told me that he asked the manager why he was removing the labels and was told, "Just do it." Apparently, even grocery stores don't understand the ways of purple-fingered stock boys.

Anyway, it's exciting to find a bunch of mystery cans. Trying to lift them can be exciting, too. Exercise care. That stock boy probably *dragged* that box out and then bench pressed it.

Often you'll find cans that are dented or bloated into interesting shapes. This may be the most shocking claim in this book, and I *know* many of you will shake your head in disbelief, but here it is: MOST OF THE TIME THE FOOD IN THESE CANS IS O.K.

I'm not talking about *meat*, naturally. NEVER eat from a bloated can containing ANY meat. That includes a can of beans with one little piece of fat. But dented cans are O.K., even if the can is bent nearly in half. It is good if it's sealed.

Most of the time I'll open a "bloated" can and find grapes, pineapple, fruit cocktail or some other fruit or fruit juice. Tasting the juice after carefully smelling the contents, I find it has a slight "kick" due to fermentation. I was told by a cannery supervisor that "low canning heat can cause on-shelf fermentation." Thus the football shape.

Anyway, most of my college-educated, germophobe friends won't believe my claim that these dented, distorted cans are safe for human consumption. It's easier to argue them out of political positions than to convince them a dented, bloated can is safe to eat from.

One of my bleeding-heart "open-minded" college roommates was particularly hard to convince. I had to bring "Scott" to my farm for a demonstration. When we arrived home, my gray-haired mother was in the kitchen cleaning a Canada goose that Jed had gunned down as it invaded our farm's sacred airspace.

"Mighty big hole in that goose," I said, sneaking up behind her.

"Well, Jedediah Bradford Hoffman, you had to use Willard's big rifle..." Mom began, then turned around.

"Hi, Mom!" I said.

"John!" she exclaimed. "And Scotty! You both here for the weekend?"

"Oh, tonight and tomorrow," I said. "Hey, I have to show something to Scotty, got any mystery cans?"

"The usual place," Mom replied, turning back to the goose. "Don't make a mess."

I opened the utility room closet and rummaged amid the cans until I found a severely dented can marked "TUNA?" I also found a small, nearly ball shaped can marked "PINEAPPLE?"

"You don't have to do this!" Scott said, looking at the cans with trepidation.

"Mom!" I yelled. "When did we get this stuff?"

Mom thought out loud for a moment, mentally placing the discovery of the cans amid known birthdays, visitations by relatives, and notable winter storms. She pinned down the date to nine months earlier.

"Nine months!" I told Scott dramatically, holding up the cans.

"Certain-fucking-DEATH!" Scott declared.

"Let's see the money," I replied.

Scott produced a twenty dollar bill.

"Don't show Ma," I said. "She hates gambling."

"You can't vomit in the next twenty-four hours," Scott said. "Or go to the hospital. Or die."

I handed him my twenty. After all, how else could he collect if I died? We shook on it.

"Mom!" I said, as we returned to the kitchen. "Got a can opener?"

Mom pointed it out amid her zillion or so utensils. Right in front of my face, of course!

I grabbed the can opener (dumpster dived behind a residence years ago), a few slices of wheat bread (courtesy of the local bakery), mayo and a bowl. Few of our bowls match. I think this produces a neat effect, very post-apocalyptic. Mom saw what I was doing and warned me that I would spoil my supper. And it was roast goose!

"I'll still be hungry," I assured her. "But I have something to prove to Scott. Hey, where did we get this mayo?"

"You know where," she smiled.

"Jerry's?" I asked.

She nodded.

"The mayo was discarded, too!" I told Scott.

He was grinning, already counting his money. *Everybody* knows that mayonnaise is practically cyanide disguised as sandwich spread. I opened the mystery can of tuna and smelled its contents carefully.

"If it's so good, why are you smelling it?" Scott demanded.

"I smell everything," I answered. "You know that. Chill."

Normally, Scott would begin a litany of my weird qualities, but my dear sweet mother was standing right there. And she was cooking his goose.

Determining the contents were good, I swiftly made myself a sandwich. Without considering the drama of the moment, I took a bite. Scott gasped out loud. So I took two more bites, consuming half the sandwich.

I opened the severely bloated "pineapple" can. But it turned out to contain fruit cocktail.

"Thought this was pineapple!" I said to Mom.

"All the pineapple is fruit cocktail. And the sardines are anchovies," she replied, testily. "Jedediah knows."

"My mistake," I answered.

I dumped the fruit cocktail in a bowl and began to eat. Scott's jaw was somewhere near his sternum.

"Want the cherries?" I asked Mom.

She *loves* the cherries in fruit cocktail.

"Thanks, honeybunch!" she said, and began picking them out with a fork.

"Duh — doesn't that stuff make you ill?" Scott asked, growing more shocked by the moment.

We often forget how strange our behavior seems to non-divers.

"THAT STUFF OUGHT TO KILL YOU!" Scott exclaimed.

"Scott," I said, looking him in the eye. "Sometimes people have deeply-held beliefs that happen to be bullshit. Like your position on firearms."

Scott was literally grabbing his hair and pulling it.

"This is not happening," he declared.

"Reality is real," I replied. "Ayn Rand."

"Praise the Lord and pass the ammunition!" Mom said.

She loves to say that every so often, though it seldom fits anywhere in a conversation.

Well, Scott observed me carefully for signs of impending death throughout supper and an evening of Clint Eastwood videos.

"Get 'em, Harry!" I'd yell every so often. "Use yer right to bear arms!"

"Praise the Lord!" Mom would chime in. "Pass the ammunition!"

Scott would shake his head as though he had accidentally beamed into the third century A.D. The next day he was bleary-eyed from lack of sleep. He later confessed that he had stayed awake all night, hoping to catch me secretly puking. When he handed over my cash, Scott boldly announced that he was ready to try a "mystery can."

"No, Scott," I said. "You better not."

"No?" he replied. "What do you mean? I knew it! IT WAS A TRICK! You must have heated the can and made it expand...!"

"It's no trick, Scott," I answered. "But you're not ready. Deep down you're still convinced the stuff is poison. You will psyche yourself into being sick."

Scott, always a sensitive guy in touch with his feelings (especially guilt and doubt), agreed I was right. In fact, he had worked up a whole rationale to explain what he had seen. Here's Scott's Crackpot Theory: The entire Hoffman family has, over the years, ingested small amounts of botulism toxin, the world's deadliest poison. Gradually we have acquired immunity. Anybody else would die from eating a bloated can of fruit cocktail. Possibly our blood could be used as a vaccine. It is worth $20 to witness such a miracle. Somebody should write to *Omni* magazine.

Yeah, right. And my cat can name all 50 states.

Plenty of cherished notions are bullshit. Once we were told by the powers-that-be to suck the venom from a snakebite. Now that's wrong — harmful, in fact. Once we were told that only homosexuals, IV drug users, and *Haitians* were at high risk for AIDS. Prostitutes were *not* considered high risk. Now prostitutes *are* high risk and Haitians have pretty much recovered from the bad press. Supposedly sugar causes cavities... but recently I read about a study that concludes it may, in fact, be starch.

So what happens to the people who sucked their snakebites, slept with whores and brushed after consuming sweets? Let them take comfort in the fact they had the "right" beliefs... God rest their souls.

Eventually, Scott was able to eat from a "mystery" can. But his whole perception of the world began to shift. Last I heard he was considering the purchase of a .38 pistol and talking about preserving his capital. I felt like the guy who had ruined Santa and the Tooth Fairy for him. It was a great feeling.

My mother and father grew up with the same beliefs about bloated cans as everyone else. So, for a few years, our pigs enjoyed lots of canned fruit. However, these animals enjoyed the stuff a little *too* much.

"Is that pig sick?" my father asked one day.

The pig was galloping about unsteadily, squealing with delight.

"Sick *nothing!*" my mother answered. "That hog is liquored up!"

Dad looked at the bloated can in his hand and suddenly thought of the "hooch" he and his buddies made in the army. They would put grape juice or some other sugary fruit juice in a can and cover it with a rubber glove. The glove would slowly inflate, then deflate, over the course of a few days. If you drank lots of it, the mildly alcoholic brew would give you a slight buzz.

"Why would stuff ferment in the can?" Dad asked. "Why doesn't *everything* ferment?"

"Maybe it's not canned right," Mom answered. "Maybe some bacteria got in."

"But the bacteria that causes stuff to ferment ain't the same as the stuff that causes food poisoning," Dad said.

They looked at the marinated hog with a sideways glance. He was standing at the trough, waiting for more peaches, grunting.

"Of course," Dad added, "hogs can eat lots of things that people can't."

"That's true, too," Mom said.

"By God," Dad said, "I bet this stuff is good."

So, right there in front of the pig pen Dad opened a can of peaches and drank the juice. We still enjoy this unusual beverage, which we call Fuzzy Hog Navel. Dad suffered no ill effects whatsoever. There was nothing gradual about this, no "building up immunity." He grabbed a can bloated into a football shape and he drank from it. Then he drank two more. And we've been eating the stuff in these "deadly" cans ever since.

Ask yourself this: if a bloated can is really full of deadly botulism, if it truly contains a toxin which can kill whole cities, then WHY aren't there special laws regarding the disposal of these deadly cans? Obviously the stuff is more deadly than PCPs, which require *years* to cause health problems. Yet no special disposal methods for bloated cans exist.

Most bloated cans do not contain botulism toxin. Of course, if a can *did* contain *clostridium botulism*, that can would probably be bloated.

What to do, what to do? Simple: test the batch. If you obtain thirty cans of fruit cocktail, feed one to a pig, goat, or other poor dumb animal. DON'T feed it to chickens. Birds are very vulnerable to even a small amount of alcohol. They will walk into walls or simply keel over, dead. DON'T open every single can and feed a portion of each to the animal. This is foolish. The stuff came from the same lot and one can will do. If the animal suffers no ill effects in 36 hours, you can be sure the can is free of botulism... a rather *rare* type of food poisoning, I should point out.

Other common food borne bacteria include *Salmonella*, *Clostridium perfringens* (involved in perfringens poisoning, which is rarely deadly), and *Staphylococcus aureus* (involved in common, generally mild staph infections).

Salmonella is killed by heating food to 140ºF for ten minutes (or less time at higher temperatures). The stuff remains alive on frozen foods.

Perfringens is everywhere. You eat it on meat every day. Only by eating an unusually large amount will you become sick. (Unless you are immunocompromised, i.e., sickly.) To control it, cool meats rapidly and refrigerate at 40ºF or below. That would include, for example, discarded veal cutlets and chicken nuggets.

Staph is controlled by heating food to 140ºF, cooling below 40ºF. However, only by cooking in a pressure cooker (240º for 30 minutes) can you destroy the toxin. That's why staph is delightfully hardy and involved in so many infections. I just have to admire it. Luckily, staph infections are rarely anything to worry about.

So, by chilling food below 40ºF and heating food above 140ºF, you are rendering it safe except for, possibly, staph. Life has risks and staph is easily controlled. When I suspect a possible staph infection in myself or a loved one I go into Mexico and buy cheap, over the counter antibiotics. No $50 doctor visit or $30 throat culture is required. I have medical coverage, but why waste my time? Mexico... what a country!

By the way, even deadly botulism toxin is destroyed by boiling for ten to twenty-five minutes. So much for the world's most deadly poison which can kill whole cities.

One health department director I know likes to hold up a bloated can of *jalapeno* peppers while he talks about the incredibly deadly nature of botulism toxin. He urges poor, underpaid food workers to toss such cans in dumpsters. (*First* he says a thimbleful can kill a whole city *then* he urges them to send the stuff to a leaky old landfill. People should be dropping dead by the millions.) If their boss gives them a problem, he says, call the health department. His government-sponsored gorillas in pinstriped suits will set matters straight. And this man *knows* about killing botulism toxin by boiling.

If those poor, underpaid workers had just a thimbleful of knowledge they'd know enough to toss those bloated cans in the trunk of their car.

I should point out that my bet with Scott was, after all, risky. We "batch test" everything but I was still risking staph. I should have cooked the stuff. But Scott knows I have a brain, and he might have gone running to a reference book. I wasn't risking death, only diarrhea. So grab those "deadly" cans and use 'em.

It's amazing how many fears our society has concerning cans. Obviously, most people find cans a profound theological mystery. I've met people who sincerely believe you'll be poisoned if you open a can and then refrigerate it. This is bullshit. This minute, as I draft this, I'm eating from a can of fried hake and salted black beans which has been sitting, open, in my fridge for two days. And the can was dented when I found it.

Dented cans cause almost as much fear as bloated cans. Plenty of people have told me the inside of a dented can will "oxidize" and make you sick. Apparently, "vegetable noodle oxide" is one of the world's most deadly poisons.

Try an experiment. Next time you are going to open a can, whack it on something and dent it.

Don't puncture it, don't break it in half, just dent it. Now open it.

Is it poison? No? Why not? Not enough time, huh? How much is enough?

So dent it half an hour before consumption. One hour. Three hours. Twelve. A day. Three days. Two weeks. A month and a half. A year. WHEN does the deadly miracle take place?

It doesn't. So long as the can isn't punctured, it is safe. Think about this: many cans are made with invisible imperfections. Why aren't people dropping dead left and right? What is the difference between one large dent and the *hundreds* of microdents inflicted on the innocent can as it travels? And what about the "flexible cans" in army MREs? Those things are treated brutally before soldiers eat them months or years later. They *should* be deadly, but they're *not*. The whole "dented can" theory is bull.

Such ignorance makes me rave but, of course, it's the same fanatical superstition that causes stuff to be discarded in the first place. So I should really be thankful.

Sweet fruits packed in water (as opposed to syrup) are the most commonly discarded bloated cans. I've noted many cans of pineapple, grapes, peaches and fruit cocktail. Many canned tomatoes, too, and tomato juice. As you open these cans, note their contents. Check for lot numbers and label all the other cans which resemble the first can. This will help you plan your meals. An indelible laundry marker works best for this, but use what's available. Don't run out and buy things.

Stack bloated cans sideways, like loaves of bread, so they don't fall over. Carefully shaking the cans will help you determine what they might contain. Kids find it fun to help label the cans. Let 'em draw a picture of a pineapple or a peach if they can't write yet. This is a delightful opportunity for "quality time." Mistakes are often made, so keep your menu flexible.

While not common, mystery cans are certainly not rare. And when you find some you will often find a *lot*. When Slash and I found a good haul we would rush in the house and yell, "Mystery cans!" It was almost like bagging a deer.

Lately, I have seen many "drink boxes," which are a delightful new trend.

Dry Goods— Things That Got Hard In The Night. Boxes and bags are wonderfully vulnerable to humidity, causing caramel corn to form a solid clump or sugary snacks to fossilize. And, of course, ugly things sometimes happen to bags and boxes. What a shame. Besides that, the things expire... especially dry cereal. And have you priced your favorite sugary cereal? How would you like to find thirty boxes in perfect condition except for having their tops ripped off?

When you find "clumped" items, just use a common sense approach. Use a grater. Dissolve things like Jell-O carefully, breaking up, the clumps with a spoon. Stale cookies and crackers can be "revitalized" in the oven. And, frankly, plenty of things that are "stale" don't taste so bad. Once I traded half a dozen boxes of stale "Crunch 'n' Munch" to a friend in exchange for a ride. This beats handing over my cash for gasoline. My friend said he would eat the stuff when he had "the munchies." He later came back and cut a deal for another two dozen boxes. He and his friends had eaten all six boxes in one night.

Don't assume everything will be stale. Most of the time these items are not stale, merely "expired." But expired goods become stale quickly, so seal the bags back up if you don't eat everything immediately. And pat yourself on the back while you save a small fortune.

Sugar and flour are found in punctured bags more often than solid clumps. Sure, the stuff is vulnerable to humidity — in your house. It doesn't stay on the shelf long enough to clump up in the store. Expired goods, besides sugary cereals, include cookies, crackers, and all manner of snack food... the stickier, the better. For some reason Jell-O, cake and pudding mixes are especially vulnerable. I can even tell you the most commonly discarded flavors: lemon cake, lime and pineapple Jell-O, pistachio pudding.

You know the little "bargain" carts in stores featuring "odd" items? If you're on a tight budget you've probably picked through these carts many times, hoping for a steal. (By the way, I notice plenty of dented cans. The management doesn't worry about a lawsuit from the family of some "fruit cocktail oxide" victim.) Ever notice how that cart sometimes becomes "cleaned out?" Well, after the manager takes his pick the unsalable stuff goes in the trash. How has the stuff "changed" twixt the cart and the dumpster's lip? It hasn't. Boy, you can find everything in these situations. Ethnic foods. Gourmet items. Kosher foods. Toiletries. Dog food. Grab it and growl.

Stuff In Bottles And Jars

Often I see the remains of broken glass containers but, naturally, I let 'em be. Contrary to popular belief, broken glass won't kill you if ingested in small amounts. But I'm not desperate.

Sometimes, however, jars become messy. One bottle of salad dressing breaks in transit and messes up the other fifteen bottles. Excellent! You weren't going to eat the label, anyway. Sometimes you'll see cans in the same condition, such as cans of chocolate syrup smeared by another can. What luck! Of course, the substances don't have to "match." You might find cans of refried beans smeared with pink dishwashing liquid. Beans tonight! Run the cans under your faucet after the meal to save yourself some dishwashing liquid.

Jars expire, but not as often as other stuff. Treat jars with "expanding" lids like a bloated can. By the way, I love those lids which read, "SAFETY BUTTON POPS UP WHEN ORIGINAL SEAL IS BROKEN." I think a few good temperature changes will pop those buttons, or maybe some leave the factory with their little buttons already popped. In any case, I find plenty of them and I'm very thankful somebody is so concerned about safety.

Be careful when you spot jars. The stock boy hurling stuff in the dumpster isn't worried about breakage. Wasteful little puke.

If a jar is only cracked, I will remove the contents and wash carefully. You can do this with large dill pickles, for example, but not a jar of salad dressing. So pour something like that through a fine mesh strainer. I won't knock myself out like this over a $1.50 jar of salad dressing, but I'll certainly do it for a quart of stuffed olives. Ever price those things? I refuse to pay that kind of money. So I dive the stuff. Exercise care and common sense and WATCH OUT FOR GLASS.

Meat — Need I Say More?

Ever held a big ol' steak in your hands and wished to God you could afford to purchase it? Heck, ever held hamburger in your hands and felt that way? Well, I can afford steaks, chops and hamburger because I save a big chunk of my food budget dumpster diving.

Steaks or hamburgers are very rare... no pun intended. But it's not so unusual to find bologna, hot dogs, breakfast sausages and other "cheap" meats. This usually means one of two things: some manager had an attack of conscience or a health inspector dropped by the store, acting fussy. Hey, maybe I called him. But I always play the game fair and only make legitimate complaints about flagrant violations.

Usually, meat which thaws accidentally is simply refrozen and allowed to become "freezer burned." People buy the stuff all the time. You have probably purchased some.

Naturally, you don't want to eat rotten meat. But you have more leeway than you think. Ever hunt deer? How long was it from the time you shot the deer until the meat was refrigerated? A day? Longer? I know plenty of people who hang a dead deer in their basement a few days to let the meat "age." So don't be a wimp. Grab that meat and get it to a cold place with all due haste. Don't run lights or act as though you're rushing a human organ somewhere — just ice the meat down as soon as you reasonably can.

Once, Slash and I found almost fifty packs of turkey franks sitting outside on a hot summer night. We dug deep, deep in our frugal pockets for some of that money we earned selling aluminum cans, selling frogs to bait stores, and independent junk sales. We pooled our change and had just enough to buy a few blocks of ice at a convenience store.

"It's worth it," I assured Jed. "This is almost a hundred bucks worth of meat!"

Jed had forked over the majority of the cash. When you're thirteen, a few bucks still seems like a lot.

"Think Dad'll pay us back?" Jed asked, hopefully.

"Jed," I told him, "If he doesn't, he doesn't. But think about this. This is *meat*. This is like bringing home a bunch of turkeys. You don't mind kicking in money for ammunition, do you?"

Dad always made us buy our own ammunition so we'd learn to be frugal. Of course, he had to purchase the ammo for us. So much for the right to bear arms.

"I guess not," Jed sighed.

"We're providing for the family," I told Jed.

"You're right!" Jed said, straightening his shoulders and putting a grim expression on his young face.

However, I ran interference for Jed. When we arrived home I cornered Dad and explained how Jed had willingly contributed to our effort, despite the fact he rarely had much money. Dad nodded and called Jed into the room. He pulled a $10 bill out of his billfold and gave it to Jed.

"You two split that according to how you bought the ice," Dad said.

"That's..." Jed hesitated. "That's three times what we paid for the ice."

"Forget it, Jedediah," Dad said. "You're saving us a bundle."

All this might seem quite weird to a child psychologist, but we thought this was a wonderful lesson in personal responsibility. We were a happy pair of divers that night.

Once, my mom found five huge steaks behind Jerry's, neatly wrapped in butcher's paper and concealed in a box. This was the first example we ever saw of employee theft using dumpsters. Over the years I have seen many more examples, often involving alcohol. These are lucky interceptions with little skill involved. But, if you dive long enough, it happens. But meat is a good source of bacteria, so be careful. Eat those steaks WELL DONE, even if you prefer rare.

Honest injun, you'll rarely find good cuts of meat in dumpsters. You will frequently find trimmings of fat and gristle. Fat is useful in making homemade soap, lard and so forth. I find all the soap I need in hotel and residential dumpsters, so why bother? You may find some good cuts for your dog, however.

Miscellaneous Items... Delightful Dumpster Surprises

Of course, grocery stores sell non-food items. Toilet paper, for example, laundry soap, tampons, brooms and mops. Drugs. Many of these things don't expire and people in the store will grab these items if damaged. Though not often, you will find this stuff. Treat with a case-by-case, common sense approach.

Once I dumpster dived behind a store that had been "sealed" for several weeks due to an IRS seizure. The store sold mostly household items, but there was candy at the cash registers for "impulse" buyers. The Infernal Revenue Service, in its finite wisdom, let all that candy go stale. (This was the same dive where I found all that Crunch 'n' Munch.) Candy is a rare find behind grocery stores, but it happens. Use in recipes or make "smores." Eat when nothing else is available. Stale candy is better than no candy at all.

Once we found several boxes of cheese in pressurized cans. The nozzles were inoperable and we couldn't get the cheese out. Bekka suggested putting the cans in a vise. We covered the cans with a few layers of towels and opened them in this manner. It was exciting and fun, a sort of cheesy bomb squad. That was the only time we used this manner of opening pressurized cans. But creative problem solving and imaginative application is useful for any dumpster diver.

Spirits are *extremely* rare. So, for that matter, is soda. Sometimes, however, you'll find a few two-liter bottles with ice crystals, or a messed up container. And you'll sometimes find a "half can" of soda. This is a can which was not filled completely at the plant. I've only observed this, so far, in soda cans. The first time I found such a can I expected it to be filled with "bad" soda, maybe even cola syrup. But it was filled with common cola.

And what a convenient serving size!

Areas behind a business are full of equipment. Stick to the dumpster and don't bother the other property.

You will often see shelving components, milk crates and other equipment near dumpster areas. Leave it be unless it has obviously been discarded. Some shelving units are so product-specific that it's tough to find any practical home use. Once we found a unit which consisted of cubbyholes the size of a bread box. We used it for our chickens to roost. But most shelving units are useless, though they appear valuable at first glance.

This is sort of weird and personal, but here goes: it gives me a rush to see commercial bric-a-brac in a down and dirty survival context. For example, when I see cardboard shacks in the Mexican *colonias,* I always feel a little rush when I see the word "Pringles," or "THIS SIDE UP," or "IBM." It's so... post-apocalyptic. So that shelving unit in the chicken coop always gave me a small charge, and I get a rush from burning wooden crates with produce trademarks stamped on the ends.

You see, commercial products are constantly hyped, creating little "recognition centers" in our heads. So, when you walk down a busy street or store aisle familiar products seem to leap at you screaming, "Buy me!" But seeing the product in a compost-splattered "no bull" context is like mental anti-toxin. You see the product leap out at you and think, "Our hogs like that!" You begin to feel layer upon layer of artificiality stripped away as you peer in dumpsters and use what you find.

Burning crates... oh, yes. Grocery store dumpsters usually contain discarded wooden produce boxes. These make excellent firewood, and you can pick these crates up by the truckload. (This makes a good cover story, too.) The main component of these boxes is often pine, so clean your chimney regularly. The boxes that are part wood, part wire are not good for burning but excellent for transporting fowl, rabbits or other small stock. You can pick up wooden bushel baskets, also, which are great for storing produce over the winter.

Hit those grocery stores regularly. They are excellent dives. Don't forget "no frills" warehouse food stores, "members only" buying clubs, natural food stores, etc. GO WHERE THE FOOD IS!

Tons of wooden crates are discarded behind grocery stores, and these are suitable for firewood. Also pictured, a shelving unit which may be useful in a basement or garage.

#3 On The Best Diving List... Bookstores

I love to read and collect books, so my listing shows a distinct bias. But books can be converted to cash or bartered, and bookstores discard a *lot* of material. Hit 'em at the end of the week for news magazines, and the end of the month for monthly magazines. Actually, the end of the month is a good time for most dumpsters, commercial and residential.

You'll find plenty of books which were heavily hyped but didn't sell so many copies. Flea market dealers will often give you ten to twenty cents apiece, but try not to dump more than a dozen copies on 'em at one time. Go to more than one dealer if you can. You're better off trading the books to the dealer for something you can use like a piece of furniture or a few good hardcover books. These people hate to part with cash but often relish a good trade.

Flea market dealers are fully aware of the "gray market" nature of these books, and many used bookstores won't touch the things. Anything with a cover is like gold, however. Take "classics" to a campus book buyer first. Once I found a whole box of classics, including stuff like *The Trojan Women*. The books were

dusty, as though they had been sitting in storage for years. I checked the titles with the book buyer on a hunch. I obtained 50¢ to a buck apiece, enough to buy two expensive texts (used, of course) for my own classes.

Lately, I've seen a few books which contain a front page saying, "WARNING: THIS BOOK IS NOT TO BE SOLD WITHOUT ITS COVER. IF COVERLESS, THIS BOOK SHOULD BE CONSIDERED STOLEN PROPERTY." Yeah, right. This fills me with the same spine-tingling fear I get from the "FBI WARNING" at the front of my videotape copies. Ooooh, I'm shaking. Rip that one lonely page out of the book and start your fireplace with it.

Don't forget to flip through the books and look for stuff like mail-in cards for cigarettes. When you notice a mail-in card in a paperback or magazine, mail in at least half a dozen for your wife, kids, dog, etc. Send to your friends, also, and people in prison for tax evasion.

So improve your mind... dive a bookstore dumpster. Good luck with all the "top three."

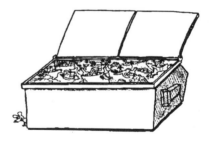

TWISTED IMAGE by Ace Backwords ©1993

Chapter 8
The "Lucky Seven" Dive Spots

Don't just dive the mall across the street. Branch out, expand, keep looking for new hot spots. Dumpster diving is like fine wine. Sure, we have one or two favorite vintages. But it's always delightful to sample something new, perhaps to find another favorite.

Discount Stores

Discount stores such as K-Mart throw out a lot of packaging and certainly aren't as hot as grocery stores, where stuff expires regularly. Discards at these stores depend almost entirely on damage. It's a diverse mix from week to week with few clear patterns. You'll find paint, cosmetics, small appliances and household items, and "broken" toys to name only a few. (You know what "broken" toys are — new toys after a week.) You may obtain potted plants, stuff suitable for building materials, even small amounts of snack food.

Those times when you really are looking for nice clean boxes, bubble wrap, styrofoam peanuts, etc., these dumpsters are a good bet. I know the owner of a small mailing service who saves money by salvaging these materials. He offered me cash to obtain these materials for

him, but his offer wasn't enough. If I had been thirteen, however, I would have jumped at it. Here's a great opportunity for kids to obtain spending money.

Frankly, it's not worth my time or dignity to tear through a dumpster looking for cans I can sell for 20¢ to 25¢ a pound. However, it's nice to know that if I were in serious financial straits I could still obtain small amounts of cash in this manner.

Having mentioned kids and discount store dumpsters, this is a good place to mention "bogus loot." As I said before, the lean and hungry dumpster diver senses the *potential* of discarded objects. Some stuff has "false potential." Kids are especially vulnerable to this. As a teenager I often grabbed stuff because it looked "neato." False teeth, for example, and used business ledgers, headless stuffed raccoons and whole boxes of empty film canisters. After a while, I would realize the stuff was basically useless.

Oh, given a decade I could probably find some clever application. But, sooner or later, even a brilliant dumpster diver will grab something that serves no purpose in the near future and only clutters his life. Food and reading material serve an obvious function, but what about a couple busted clock radios? You can cut the cords off and sell by the pound to a scrap dealer. You can give the clock radios to your genius child for his amusement. You might fix the things or trade to a small appliance repairman. (Talented people like this are as rare as talking dolphins.) So, even though a broken clock radio may seem "neato" at first glance, the applications are actually quite limited.

Slash and I made a snowmobile sled from the hood of a '56 Chevy, insulated the walls of our hog house with bubble wrap, and used department store dummies to drive the local swine patrol insane. But even innovative lads like Slash and me would sometimes grab something useless like a giant cardboard Easter Bunny or a crate of empty aerosol cans. So we'd burn the stuff to keep warm, shoot holes in it or discard it. Discount stores often contain "neato" pieces of store displays. But the only thing giant cardboard figures are good for is decorating a kiddie clubhouse or starting your fireplace. A word to the wise.

Here's a handy tip: Want to provide your little ones with hours of fun and save yourself a bundle? Give your kids a giant cardboard box at playtime. Kids love big boxes. Bubble wrap with giant bubbles is fun, too, but don't let them wrap it around their heads. Slice the stuff in manageable chunks. People are warned constantly that plastic bags will suffocate children, damage their brain cells, etc. If you're worried about your child's brain cells then yank 'em out of those damned public schools. THAT'S brain damage. THAT turns people into vegetables.

Teach your kids how to dumpster dive, how to *do* things. Go to the discount stores and obtain the best discount of all — GET IT FREE!

Candy Stores — A Sweet Deal

A little candy goes a long way, making candy stores an extremely good dive. For really great hauls, dive these places after Christmas, Easter and Halloween. Valentine's Day has potential, too. The stores can't sell *all* those heart, bunny and Santa-shaped candies at 50% off.

One problem I often encounter is that plastic wrappers will adhere stubbornly to sticky candy. Attempting to peel this stuff off with your fingernails is time consuming if you are making, say, caramel apples. I've had good luck dipping this stuff in boiling water for several seconds and then scraping the melted candy off with a small spatula. Hold the rubber end of the spatula in your hands like a stone tool for better control. Kids can often be enlisted in the tedious process of peeling the packaging off sticky candy. If they are little and can't help themselves, let them eat as much as they like and make the candy the *next* day. If you have a dog WATCH OUT when doing sorting of this type. They'll dive into that pile of sticky plastic and *graze*. But remember that a little piece of paper or plastic in your food won't kill you.

For a great dumpster treat, toast stale marshmallows and sandwich them between revitalized cookies or graham crackers with small pieces of stale chocolate. This chocolate will taste relatively fresh when melted, so break in pieces small enough to melt instantly when you slap that hot marshmallow between the cookies. Burn pine crates with apple or peach scent for atmosphere. Read a salvaged kiddie book to your children or work on an artsy project with dumpster dived materials. This is real quality time.

Seldom do I think to myself, "Why, I dumpster dived everything at this table!" Nor do I set out to create all-dumpster projects, meals or decor. Rather, I smoothly incorporate dumpster goodies into my life. But sometimes I do look around and say, "Wow... I've got a lot of great stuff!" I look in my wallet and note that extra ten or twenty. I think for a moment where I obtained the shirt I'm wearing. And, while I don't have any children, I know it made my parents feel good to watch me, Bekka and Jed eating fresh fruit, meat and candy that would normally cost a limb and a vital organ.

The other day I watched a woman on television bitching about the bad economy. She said it broke her heart when her kids asked for candy and she had none to give. Only the previous week I had found enough butterscotch candy to last three months, not to mention two dozen rolls of Lifesavers. The lady on television lived only a mile from the dumpster in question.

"Tsk, tsk," I said.

This wasn't pity or sarcasm... I was sucking on a piece of butterscotch.

Often you'll find bulk candy discarded in cardboard boxes, all mixed together. Put that candy in an airtight container even if it's stale. There are, after all, levels of staleness. If you have a large amount of candy a clean plastic trash bag can work, too. Enjoy in moderation.

Wholesale Florist Dumpsters

*Wholesale florists will allow you
to carry on multiple romances.*

Small florists buy their blooms from wholesale florists, who receive their goods from growers and/or middlemen. Flowers are even more perishable than fruits or vegetables. Plenty of stores will sell a small or oddly-shaped

cucumber, and plenty of customers will buy one, despite the paranoia about "perfection." But would you purchase a blemished long-stemmed red rose? A less-than-fresh carnation? The dumpsters behind this business are *bursting* with ever-so-slightly wilted roses, smaller-than-average carnations, depressed daises and other flora. You will often find flowerpots, vines, potted plants, and in some cases even seeds and small ready-to-plant bushes and trees. Raid those large nurseries and florist shops, too. A dumpster diver's table should always contain a lovely floral arrangement.

Of course, clever dumpster diving dudes interested in the opposite sex will quickly grasp the concept of converting flowers into romance. For example, the day after Valentine's Day, I, a lowly college "frosh," presented a lovely senior lady with 144 long-stemmed red roses. That's a dozen dozen.

We've obtained so many small trees from nurseries that we have planted a virtual forest in our rural township. After we filled our orchard with trees we planted isolated spots around our favorite hunting areas. Naturally, these baby trees are in bad shape, dried out or damaged by rough handling. About 25% survive after planting, even with care. But have you ever tried to *buy* fruit trees? Or flowering bushes? Plant enough fruit trees and you may be able to give up dumpster diving. Not that you would *want* to, of course.

Naturally, the days before and after Arbor Day are hot around tree nurseries.

Ornamental vines and potted plants are frequently discarded. These require only a bit of tender loving care to flourish. It's hard to believe "plant people" would do these horrible things to plants, but plenty of people work in jobs they actually *hate*. You should see the books I find behind libraries and the dead animals behind pet stores. *Sick.* Post-dumpster survival of potted plants varies considerably, but 80% is about average. You can find almost as many potted plants in residential areas.

This is just my view, but I hate to see living plants in dumpsters. Things that grow from the earth should go back to the earth, not to the landfill with all the plastic and chemicals. So, whenever I see a plant I grab it. If I don't *want* the plant, can't trade it and don't feel it can be saved, I plant it on a vacant lot. Once I returned to find a wilted, sad little vine had virtually colonized the lot. This gives me a feeling of oneness with the earth.

By the way, what good are a hundred wilted roses? Make potpourri. By the way, the flowers you obtain will last longer if you trim the stems before putting them in water.

Toy and Novelty Stores

There's more to be found in these places than heart-shaped handcuffs. Toys are obscenely expensive, and even broken toys are plenty of fun. After a toy machine-gun no longer makes a pop! pop! sound, it is *still* a useful toy. Toy trucks that no longer spin their wheels are hours of fun in the sandpile, anyway. Do yourself and your children a favor — stop buying new toys and *feel good*. Dive toys in dumpsters! Don't be a slave to a Madison Avenue guilt trip.

Toy stores are *wonderful* places from a dumpster diving perspective. There are always "display" toys being handled in a brutal manner, boxes torn open by eager little hands, heavily hyped items selling poorly in a subjective and emotion-charged atmosphere. Of course, plenty of toys and novelties have a limited shelf life — such as Halloween costumes and toy Easter bunnies. Toys are seldom sturdy and frequently composed of little parts which disappear into an alternate universe. A dumpster diver who consistently dives these dumpsters will eventually acquire enough bike parts to construct a ten speed. Certainly, you will find a fixable trike for your little one. Unlike food, a few toys go a long way.

Use a common sense approach, and get your children involved in the process. If you see a teddy bear losing his stuffing, sew him up. My mom would don a pretend surgical mask and let Bekka and me watch as she "operated" on injured stuffed animals. She was a miraculous surgeon, reattaching limbs and eyes good as new. The "patient" would be conscious the whole time, numbed with pretend medicine, commenting hopefully on the procedure.

Be careful when you acquire these toys, or ANY toys for that matter. Look for sharp edges, loose parts, and any other possible hazard. And remember, kids see toys differently than adults. My GI Joe was a real American hero, because he never allowed his missing arm to affect his life. To the horror of their parents, the Kietzer twins modified their new GI Joe figures accordingly. Tea sets don't have to match, toys can be dented or scratched and still be *fun*. Little Jed had so many toy guns that we called his room "the armory." When his (weird) little buddies came over to play, they had several guns apiece and *real* pieces of army uniforms. Playtime would culminate with the firing of actual rifles under the close supervision of my dad.

Liberate yourself from an expensive guilt trip — dive those toy store dumpsters for your kids or to acquire salable items for yourself.

Restaurants, Fast Food — Down And Dirty Diving Opportunities

So, let's just say it: dumpsters at food places are somewhat disgusting. But the payoff can be big.

Think fast food. Burger places and fried chicken places make food which sits on the hot rack, awaiting customers. After twenty minutes to half an hour it is discarded. Pizza places mess up an order, then toss it in the trash.

Here's the key at burger and chicken places: find out what that bag under the front counter looks like. Plenty of places use a small white bag in the front and big black bags in the back. Use any ruse to get a look at that bag. Peer through the door that leads behind the counter. Roll your child's toy ball behind the counter and rush around, apologetically, to retrieve it. Get a look at the bag!

Fast food places with hot racks have three distinct categories of garbage: A.) Customer crap. All those empty drink cups, half-eaten burgers, french fry boxes, etc. Too much sorting unless your hogs are *starving*. B.) Kitchen crap. Mountains of coffee grounds, plastic pouches of shake mix, cardboard containers, etc. Sometimes you'll get lucky and find a bag of buns or something. C.) HOT RACK TREASURE! Comprising roughly 5% of the discarded stuff, this is what you seek.

It's not impossible to find, but it's not easy. Your task is made harder by the security fences many fast food places have erected. Explaining your presence inside a security fence is no picnic. So check out every fast food place until you find one without a security fence, or with an unsecured entry like a simple latch. That dumpster is your baby.

All you have to do is *find that bag*. Visualize it. It's slightly heavier than most bags in relation to its degree of "fullness!" It contains boxes and boxes of burgers, loose french fries, maybe a coffee filter or discarded straw wrappers. *Believe* in the bag.

Once you locate it, the rest is easy. Toss those burgers in the microwave or warm them in the oven. Scrape off lettuce and tomatoes, as these veggies become limp and tasteless. Replace with fresh dumpster dived toppings. Try to salvage some of the fries by carefully lifting them out of the fry boxes, salvaging what you can. Fried chicken is usually dried out and turns to leather with rewarming. Scrape the meat (including the breading) off the bones and make cream of chicken. Serve over dumpster dived toast. Delicious!

When you become proficient you'll have a surplus of burgers or chicken. Salvage the meat, put it in bags and freeze for a rainy day.

Or a snowy day. Reminds me of a story. As a freshman I used to do a lot of my homework at a local burger place. They had a "free refills" policy on soda and coffee, so I would just sit there all day with my study partners and do my homework. Well, one day a severe winter storm hit the state like a big white wall. Businesses in "College Town" closed early so people could make their way home. I watched a steady stream of customers flock through the burger place — then business became dead abruptly as the storm hit.

"What are you staring at?" Rhoda, a rich friend, asked.

I was staring at the hot rack, transfixed.

"All that *food!*" I answered.

Right then an employee came along and began to clear the hot rack. I watched him toss dozens of burgers in a plastic bag, then grab the bag and disappear with it.

"Gimme your car keys!" I said to Rhoda.

She handed me the keys with a quizzical look. I ran out the door and saw the employee toss the heavy bag in the dumpster. That bag wasn't it in the dumpster more than a few seconds when I ran up and snatched it. I made my way through the blowing snow to Rhoda's car and locked the bag in the trunk.

"Let's go home!" I said, going back inside.

My dorm was just across the street. We could walk across and leave the car until the storm blew out of town.

"Why?" Rhoda asked.

"I found something!" I said.

Rhoda knew what that meant. And she quickly connected my disappearance to the empty food rack.

"You went and grabbed those burgers from the *trash?*" she hissed.

I nodded vigorously. She smiled.

"You going to give those burgers to your friends under the bridge?" she asked.

"What friends?" I answered.

"You mean you're going to eat that stuff?" she asked. "What about the homeless people who dig through that dumpster?"

"Do they come knocking on my door offering to share with me?" I asked.

"But ...!" she was aghast with upperclass guilt. "But you have more than they do!"

"Does your father come knocking on my door, offering to share with me?"

"That's different!" she said. "How will the world become better if we don't help people who can't provide for themselves?"

"Are you going to keep spreading Marxism or are you going to help me eat some burgers?"

Rhoda considered this for a moment. She could probably pull out her checkbook and *buy* that burger place, but she was always chasing down new insights among the downtrodden. Like many people who inherit large amounts of money, Rhoda was searching for a *connection* with the so-called "real" world. Whenever she gained some sort of working class insight she would write down her thoughts on expensive stationary, keeping the papers in an attractive leather portfolio. Once, against her instructions, I peeked in the notebook. I wanted to see if she was learning anything from me.

"Slopping hogs and plucking chickens," I read, in an elegant, flowing hand.

"*This* will help her run daddy's empire," I thought.

Because Rhoda considered me a source of rare experiences, she treated me rather like a person from a primitive culture: if my customs offended

her, she was willing to endure it in exchange for a learning experience.

"I'll help you eat the burgers," she said, packing up her books.

And, as it turned out, she ate more than me.

Burger places are an area where "interception" can be an important factor. During my senior year of college "Scott" worked at that same burger place. When he carried out the "hot" bag he would mark it with an indelible marker and toss it in the upper right corner. His clever little "Burger Boyz" costume had no pockets, so Scott carried the laundry marker in his sock.

Late at night I would raid the dumpster and grab the bag. If I had been caught and forcibly detained I would have kept my mouth shut about Scott's role. Scott and I would split the burgers and have a feast. Every few weeks I would bring home frozen bags of hamburger, fish and chicken to my parents. Scott would use the money he saved on food to purchase beer.

This basic "interception" method, with subtle variations, can be used at any business. The key is to *be the one who takes out the trash.* Jump in and lend a hand until that's *your* job. When my brother worked as a grocery store stock boy (oh, the irony!) he would use this same trick. When he had a huge stash, or knew of one, he'd call home and talk to a family member. "How's Elmer?" meant a personalized delivery in the right corner. "How's Bipper Bunny?" meant we'd have to hop all over to find the stash. Asking about both pets, of course, meant both types of loot.

This simple system worked to perfection and never grew more elaborate. The manager, by the way, was very impressed with Jed's willingness to keep the stock rotated and his touching concern for his pets. Jed was voted Employee of the Month and Employee of the Quarter, receiving $50 and a plaque both times.

Jed might have made Employee of the Year if our family had ever purchased groceries. The manager was always lamenting the fact Jed's family never came by the store. Little did he know Jed was sometimes flinging boxes of chow right in our hands, acting like he didn't know us. Jed kept the stock rotated *daily*. A dented can or expired milk carton was not to be found anywhere in the store on Jed's watch.

Back to dumpsters behind food places. Go to quality restaurants and grab the bags that make "squishy" sounds or feel extremely heavy. The places don't need to be expensive *so long as they don't use disposable cups and tableware*. Such places produce C-R-A-P. It's not worth the time it takes to sort unless, as I said before, your hogs are starving and no alternative is available.

WATCH OUT for broken plates and glasses. Pick through the hog chow with thick gloves. Smash up the broken chinaware (NOT glass) and toss it around for chickens. This aids their digestion. Pull out the large, jagged pieces and don't sweat the small stuff you can't see.

But DON'T allow hogs to eat raw pork. This can give healthy hogs more problems with worms than they currently experience.

I feel so much better about slaughtering hogs when I know the animals had a short, happy life feasting on oysters Rockefeller, pizza, pasta, chicken, burgers and all manner of produce. And it gives me a cheap thrill to watch our hogs eating sirloin and shrimp while people on t.v. bitch about their grocery bill. We give the porkers names like "Taco John," "Sheraton the Swine" and "Angelo" so they remember who is buttering their dinner rolls.

In regard to HUMAN chow: eateries don't throw out "good" food like grocery stores. A store can't sell bruised, soft melons but a restaurant will trim and utilize "bad" produce without blinking. But plenty of good things can be obtained if you're there at the right moment. Brown bananas, for example, and broccoli stalks. Bread ends, neatly wrapped in plastic bags. And lots and lots of stale dinner rolls.

Check these places after a major power outage. (One that *lasts* a while, as opposed to one

that is widespread.) Grocery stores usually have better emergency measures because of their huge, valuable inventories. I swear, I have found *filet mignon* like this, five gallon buckets of ice cream, and other goodies too numerous to name. But the best approach with eating places is to hit the fast food joints and look for the "hot bag."

Behind restaurants, you'll find plenty of discarded five gallon buckets. These are containers for such goodies as salad dressing, tartar sauce and other condiments, various syrups, even pickles. The buckets are sturdy and fantastically useful on a farm. I've seen them sold for a buck apiece at flea markets.

Collect the buckets and use a spatula to scrape out a pint or so of salad dressing from each bucket. Wash and use the buckets. I like to leave the labels on the buckets and particularly like the ones with brand names stamped on the plastic. But that's me. Many of the buckets will retain the aroma of their original contents even after careful washing. I consider this a bonus. These buckets are excellent for long-term storage needs. Remember to grab the covers, too. You can also find many produce crates behind restaurants.

Quality restaurants and hotels which use "chafing dishes" at banquet functions throw out lots of sterno containers, as well as other types of "canned heat." Frequently, these are only half-used. You can use "as is" or take out the fuel. Drain the ones which contain alcohol or extract the solid "canned heat." The waxy blue canned heat is so pretty that Slash and I just hold it up to the light and say, "Wow...!" Don't smoke while you do this.

If homeless people are in the area, they will compete vigorously for this canned heat. The stuff is a great survivalist item, and if I were homeless I'd find a big hotel and dive for the stuff. As it is, I've used the stuff on my adolescent bike trips.

Eatery dumpsters are messy, but with luck and skill you can experience excellent pay-offs.

Residential Areas — The Ultimate Dive

You haven't really experienced dumpster thrills until you've dived residential. Other dumpsters are more reliable, or produce bigger bonanzas, but for sheer variety and fun you can't beat thy neighbor's dumpster.

Ultimately, everything in our society is produced for consumers. And, sooner or later, most of this stuff is discarded. Even things made for specialized business applications — like uniforms, copies of *NASA Tech Briefs*, and sets of law books — end up in residential dumpsters. Sometimes I am awed at the incredible journey these items made before ending up in my eager hands: for example, Chilean abalone canned in Hong Kong, sold in San Francisco (judging by the receipts) and then discarded on the U.S.-Mexico border.

Plenty of stuff in residential areas can scarcely be found anywhere else. Clothing, for example. I haven't mentioned clothing stores because these places seldom discard anything but boxes and hangers. If they want to dispose of unsaleable clothing they sell it back to bulk clothing buyers. If you want clothing — go residential.

These dumpsters are delightfully personal. You'll find mail, private personal papers, photos, discarded credit cards, unused checks and ID, books, magazines, furniture and household items. You'll find an amazing amount of usable food. You will find this week's copy of *Newsweek*, potted plants, working appliances and whole boxes of half-used toiletry items.

And, of course, plenty of crapola. Kitty litter. Used diapers. Snot filled tissues. Empty cans, cartons, and pizza boxes. But that's no big deal thanks to "trash segregation." A box of paperbacks might be right next to a bag of disgusting crap... but the twain shall never meet. Buck up and dive.

Middle class neighborhoods are best. Working class and upper class neighborhoods can be good dives, if the working class neighborhoods are tidy and the upper class neighborhoods lack xenophobic security forces. Get your classes straight, too. Plenty of working class stiffs fancy that they are "middle class" just because they made cannery line supervisor. Plenty of upper class people who own controlling interest in a bank fancy they are "just middle class folks trying to make a living." Yeah, gimme a break.

Working people discard a surprising amount of good stuff, but don't bother with crappy-looking neighborhoods unless the dumpster is on your way. Rich people throw out surprisingly little good stuff, but it was probably frugality which made them upper class. What you do find is worth the effort. This is a glimpse at the "lifestyles of the rich and famous" which few people ever see. Does she or doesn't she? Only the dumpster diver knows for sure. (Answer: She does. And often.)

Quality and content vary wildly from week to week, neighborhood to neighborhood, dumpster to dumpster, even from bag to bag. Currently, I live in a neighborhood with lots of military people. I frequently find specialized army manuals, pieces of gear, even MREs. A neighborhood only a few blocks away is full of retired people. As these old folks die I find piles of discarded "junk," including 1920s costume jewelry, old cans, vintage clothing and interesting documents. Some junk! Many people have paid good prices for my dumpster pickings in some high-priced places.

The best time to dive residential dumpsters is when people are moving out. When you spot a moving van, start skulking. Frequently, however, people are moving with no visible indications that this is taking place. You'll know when you find the stuff in the dumpsters. Here's a tip: when you find discarded mops and brooms, *tear open every bag in the dumpster.* Plenty of people moving from a rental facility will buy and then discard mops and brooms. Often, it's tough to tell the difference between moving, and ambitious cleaning/remodeling projects or death (which is a kind of move). Other things like a job change, child leaving home or change in marital status produce a lot of good junk. Every day people make little changes that produce good discards and interesting trash.

Once, I found so much good stuff and so many highly personal articles that I wondered, "Did this young lady die?" The stuff included photograph albums and diaries dating back to the age of seven. There were more than a few sexually explicit photos and old appointment slips for an abortion clinic. I figured out very quickly that this young lady was a self-critical over-achiever who dated abusive young men. There was even one old letter which began, "I know you're wondering why the return address on the letter says, 'Steve.' Well, the answer is that my grandmother is going to let me stay here and go to school and pretend I'm Steve until every-thing cools off back in D--- County. So how are you? Sorry I didn't write or anything but I was kind of keeping a low profile."

Fascinating stuff. I looked through it for hours, drinking some fine Earl Grey tea I discovered in the top part of the box. No, the young lady had not died. But I found out she had made a serious suicide attempt only days after discarding most of her life (including several books of blank checks!).

Now, if I knew thousands of intimate details about a depressed, confused young lady WHAT could I do with that information? Many, many times I have found out the damnedest things digging through somebody's trash and won-dered, "What could I do with this?" A more ruthless person would pose as a dashing young man with psychic abilities. A more ruthless per-son would urge his wife or girlfriend to dye her hair, get some glasses and start passing around a *lot* of bad paper.

I'm not *that* ruthless. But it amazes me that people seem to think tossing stuff in a dumpster is some final act, like hurling it in a volcano, as though human eyes will never glimpse the stuff again. Many times *I* glimpse those wild little secrets and wonder to myself what I could do with a Mexican voter's card, an Electric Com-pany identity badge, and so forth. I even found some sort of Satanic get-up one time, and a

pouch from a bank another time. Some really great practical jokes begin to materialize in my mind when I find such things.

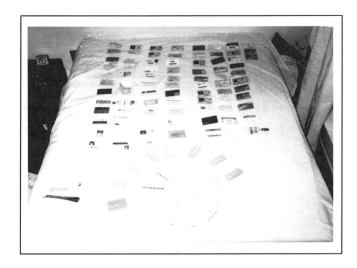

Dumpster dived forms of ID, credit cards and checks dived over the course of a few months.

Skulk big, nice apartment complexes with high turnover. My own apartment complex is full of military officers and the lease has a liberal military "escape" clause. The city is a "buyer's market" for housing, with people moving often. That's hog heaven.

The process of moving mysteriously trans-forms people into idiots. They take everything in their fridge and toss it out. They toss out the stuff in their pantries. They throw out whole boxes of soap, shampoo and medicine, some of it never opened. Finding neat stuff like Valium is rare, but it happens. You can find unopened rolls of bathroom tissue and even boxes of un-used diapers. I've found dinette sets missing one chair, spice racks missing one or two jars of spice, couches with a ripped cushion, flashlights with dead batteries, you name it. Half the clothing I find comes in bags or boxes *washed and folded*. This never fails to amaze me. People throw out underwear without taking it out of the plastic bag. People throw out *lots* of books, including a surprising amount of erotic material. That includes everything from deadly dull copies of *Playboy* featuring LaToya Jackson to specialized cross-racial fetish mags.

Residents often throw away valid credit cards. These cards often have one month left before expiration, so I conclude the individual just received a new card. Forms of ID can be found frequently in discarded boxes of "odds and ends." I believe crime doesn't pay, but I often find myself utilizing a fake "student discount," park pass or other small perks. And, of course, I can't help saving forms of ID, blank checks and credit cards. It's nice to have something to fall back upon in an uncertain world. I've found every form of employee ID imaginable, passports, drivers licenses, medical insurance cards, you name it. I've found plenty of credit cards clipped neatly in half, plenty of credit card receipts, typeset resumes, completed forms filled with personal information, letters filled with interesting tidbits, and so forth. Many of these belong to dead people.

I admit it... I'm a snoop! I like to pick up a "story box" and figure out things about people. Trashy stories can be found everywhere. I call this "ersatz archeology." I love to find out about the troubled young woman discharged from the Navy for psychiatric problems... the compulsive dieter, bingeing once again... the home based business that went bankrupt... the woman who had a baby, lost her figure, and tossed out all her college textbooks... and much, much weirder stuff. What is that pair of ballerina slippers doing with the child porno mags? Why would somebody throw away a wad of banknotes from Trinidad? Just how many penpals does this man have? And why do they always talk about *feet*? And, much, much more. Garbage is intensely human and personal, terribly interesting stuff.

Besides being a snoop, I'm a nut for specialized magazines aimed at different professions. I love to find copies of *Restaurants and Institutions* because I know I can send off the mail-in card and obtain, perhaps, ten pounds of muffin mix or a can of vegetable oil.

But I also read the *American Bar Association Journal, Physics Today, Police, The Pipeliner, Journal of the American Medical Association*, and every annual report I can obtain. And I learn the damnedest things! This is so much more conven-

ient than using the stuff at the library, and more serendipitous, as it were. Sometimes I find a whole pile of material that *doesn't* interest me, but it may be gold to you. Sometimes I feel SO strongly about the things I read that I write letters to the editor. Hard to tell if they get published — I won't see another issue unless somebody discards one.

Hardcover books and paperbacks are gold to me, even if the topics don't interest me. I can trade copies of romance novels for copies of *Black Powder Digest, Finding and Buying Your Place In the Country, The Seven Laws of Money*, and so forth. You may prefer to trade those titles for romance novels. Whatever.

Slash and I once tried to pinpoint good dumpster dives by reading the obituaries. It didn't work. Stuff can be tied up in probate for months before somebody gets control and promptly discards half the stuff. But if you know an old recluse in your neighborhood knock-knock-knocking on heaven's door, keep your eyes peeled.

Once Jed and I found several boxes of old school books from the 1800s and a load of "vintage" clothing. Sometimes you'll spot a lot of good stuff, load it up, and later figure out that the resident must have died. All the more reason to move with efficient haste and to avoid verbal confrontations.

Residential dumpsters are full of aluminum beverage cans. People drink more beer and soda than you can imagine. Often you'll get lucky and find an aluminum stepladder, screen door or other aluminum scrap. And, of course, whole families on the south side of the border make a living selling aluminum cans. Nowadays, can collection is *not* worth my time, but I keep abreast of developments. Besides aluminum, dumpsters are full of copper in the form of electrical cords. You can sell these by the pound to scrap dealers.

It's lovely to watch the change of seasons and special events in dumpsters. Around Thanksgiving, you'll find half-eaten turkeys, perfect for your pet when you pick off the meat. Christmas means the piney scent of discarded trees and tons of wrapping paper. Wedding showers are followed by baby showers and dirty diapers. Clothes are discarded as the seasons change. People buy new things and throw out old things. Children outgrow their clothes and toys. Potted plants wilt under neglect. People die, people move, styles change, technology moves forward despite the government, and gradually people acquire more elaborate televisions. Always, people eat and throw out tons of packaging crap. But sometimes they get confused and throw out the food *before* they've eaten it. Every now and then somebody throws out something really good.

Toys picked up in an apartment complex in a single dive.

As I said before, I'm picky about residential food. It's been through somebody's hands twice, double the danger of store-discarded food. But when I find something like a freezer-burned ten pound turkey, a three pound slab of cheese or a hunk of smoked sausage I grab it *joyfully.* I always grab stuff in cans. Be picky, not paranoid. Plenty of people throw out good food for dumb reasons.

Use a common sense approach with discarded toiletries, pills, detergents, etc. For example, if you find a half-used roll-on deodorant, carefully slice off the exposed "used" portion. Save and use the rest. Clean the end of toothpaste tubes

and squeeze out a bit of paste before using. Use bottles of shampoo and boxes of detergent "as is." Handle medications on a case-by-case basis. Remember, if it's sealed, it's *good.* If it's unexposed to the elements, it's good, too. But beware of expiration dates with such fragile meds as antibiotics. Some stuff *can* turn on you. In the case of pills in open containers, carefully examine them before using. After all, you'd use a Tylenol from a friend, wouldn't you? Think of residents as your friends, your clean and friendly pill-dispensing friends.

The problem with most medications isn't contamination but *usefulness.* Painkillers and mood-altering substances aren't as common as vitamins and diet supplements. Once I had over thirty pounds of assorted vitamins. Jed would run his hands through the box and do his Time-Warped-Pirate-Turns-Narco-Smuggler routine. Finally, we ground up the vitamins and tried to fertilize some plants... just as an experiment. The mixture looked like powdery vomit. The plants died within a day.

A physician lived in my neighborhood and he received many "samples" of various medications. I used to find whole boxes of cough syrup samples, even strange foreign meds. This dumpster was so dependable I began to call it the "Med Box." After a while the doctor relocated and a chronic dieter moved into the vacant apartment. Whenever she went on a diet she would throw out most of the food in her house. I would find unopened boxes of snack food, frozen entrees, half-melted fudge bars, piles of frozen goodies that reminded me of a supermarket dumpster. Her struggle with food was evident by such clues as writing on the containers. Some boxes had the words "NO WAY!" A tightly sealed jar with electrical tape on the lid had the words "REAL SUGAR" written on the front. I once found a whole set of tapes from a nationally-known diet guru. The boxed set of sixteen tapes (minus two of the "binge" tapes) cost, I'm told, about sixty bucks. My wife listened to some of the tapes, found them useful, and promptly lost fifteen pounds.

Like any professional, dumpster divers develop their own jargon. Jed will tell me, "Hey, let's hit that dumpster where we found all the carnations." Pretty soon this is shortened to "carnation dumpster." A more spectacular find at a particular location will change or modify the name. Sometimes we'll talk about our "raids" or "hits" and our "E.T.A." Don't take yourself too seriously and don't be afraid to have fun.

Residential dumpsters are so varied in their content that considerable smarts are needed to use what you obtain. A good trade network can help a lot. You may not have any need for blank prescription pads but this stuff could be extremely valuable in a barter situation.

Dumpster Diving and Higher Education

I've dumpster dived in campus areas both as a student and a "townie." In fact, dumpster diving made a vital difference to me while I sought my degree. Some other dumpster diver may have a *magna cum laude* but I've never met him or her. More proof that "dumpster diving is like crime." It's not that *crime* is dumb, just that most *criminals* are dumb. When I tell you that dumpster diving can help you achieve your goals, I've been there and achieved those goals.

As a college student I never bought a pair of shoes or other clothing except, a few times, at rummage sales or second hand stores. I never bought a newspaper in four years of college. I did buy typing paper a few times when I couldn't "liberate" some. However, I never bought a notebook, pens, paper, paperclips, staples, or numerous other items. Frequently, I had a small "stockpile" which I could sell or trade for other items. I even dumpster dived sheets of fabric softener which I reused to keep my laundry "sunshiny fresh."

Let's examine college kids. Lots of college kids are wasteful little pukoids. Plenty of them have never worked for a living and don't realize the value of possessions. They'll cry about the poor hungry people in Ethiopia, but watch 'em throw out a whole fridge full of good food.

Obtain class schedules at your local campus and *know when the semesters end*. Moving Day (when the dorms must be vacated) is a few days later. Spring and Fall Break as well as the beginning of Christmas Vacation are hot, hot times. You can pick up furniture by the *truckload*. Funny thing is this: you can sell the furniture back to *other* college students.

You can also pick up textbooks, though only ten percent of the books can be resold to a textbook buyer. However, I've made $100 cash money for twenty minutes work simply by raiding the trash cans of a dorm full of Arab students. Even textbooks which have been subsequently updated (and the old edition can't be resold to a textbook buyer) are useful and worth a small amount to a used book buyer. Of course, you may find it useful to have your own textbooks on investment management, biology, Native American studies, higher mathematics, etc.

Here's a hot tip: walk into the various campus departments and look for "give away" textbooks. The companies which print textbooks send sample copies to profs, and these profs often give away or discard the samples at the end of the semester. These books may have the word "sample" on the front in gold letters, or a sticker on the cover. Alter before selling to a textbook buyer.

College kids have some unique habits which show up in their trash. For example, they are usually not allowed to have dogs or cats in dorms. However, they use their overly-large allowances to obtain such exotic pets as snakes, lizards or tropical fish. Being irresponsible, first-time snake owners, their pets often perish. Plenty of the little creeps just flush the fish at semester's end. You can find lots of aquariums, fish food, little nets, pretty stones, etc. Even a cracked aquarium makes a good terrarium with a little work.

Slash obtains damaged aquariums, patches them up, and creates nice little terrariums. When Slash visits me in Texas he acquires a bunch of tarantulas, free. Back in Minnesota he sells the tarantulas and aquariums *cum* terrariums for $30 or more. CAREFULLY CLEAN all aquariums lest your critter die from the same thing that killed the *other* critter. Be careful, by the way, with bird cages, animal travel compartments and things like dog dishes. Protect your pet.

College kids use space-saving "lofts" also. At the end of the semester you can pick up next-to-new lumber, bolts and nuts from discarded lofts. You can also find plenty of cinderblocks pressed into service as shelving components, sturdy milk crates and carpeting remnants.

College kids love to collect some weird stuff. Usually these are items acquired during a night of drunken debauchery or in a scavenger hunt that went out of control. Road signs, hazard lights and full-size "burger dude" figures, for example. Slash loves to collect signs, and has decorated his bedroom and several utility buildings in this manner.

"Wild" Willard Hoffman once made good use of a couple signs acquired on a campus. Despite repeated appeals to our city, Hoffmanville couldn't see the wisdom of installing "WATCH FOR CHILDREN" signs in our isolated neighborhood. Documented incidents couldn't sway the paper-pushing bastards. Then Dad had the incredible luck of acquiring two such signs on campus. He found a couple metal posts left in a ditch and established his own 30 m.p.h. zone.

Years later the signs were growing rusted and had a few stray bullet holes. Dad boldly asked the city council to replace the signs. When the flustered red-tapists couldn't find the "proper papers" on the signs, Dad raised hell about their idiocy. An individual campaigning for local office gladly cited their incompetence, too.

A week after the election we had our new signs.

Once, I called a business and offered to return their "burger dude" figure... for a price. The manager acted like a total asshole and implied that I might very well be responsible for the "hamburglary" in the first place. Luckily, I had provided a fake name and address. So I hung up on the bun-stuffing bastard. A week later the figure was found on a public median, wearing a Mexican sombrero and women's underwear. A sign hung around his neck. "My meat hurts!" A tragic incident that could have been avoided.

The signs and figures you find in campus dumpsters often have an ironic twist. For example, "CLIMAX, MN" has a helluva time keeping its sign. "SLIPPERY WHEN WET" and "MEN AT WORK" signs were popular a few years back because of musical associations.

Well, you can't sell the signs for scrap. And the victimized town or business will probably be less-than-grateful if you offer a return. The stuff is "neato," but often useless.

So, generally, campus areas are a great place to pick up a large amount of books, notebooks, light household items and other select materials during certain "hot" periods. You can often find essays which can be sold to *other* college students. And you'll find a surprising amount of checks, credit cards and forms of ID. It's useful to have a campus ID so you can attend cultural events and receive student discounts. I rarely find less than a dozen forms of ID and at least one book of checks at the end of the semester. If you are "inside" the campus then odd papers from the office of the dean and an essay or two are *very* useful.

By the way, watch the college library closely. For that matter, most libraries are a good bet. As books become worn, damaged or grow old these volumes are routinely discarded. Nice old magazines, too! I once found a huge box of *National Geographic*. I was able to extend my personal collection of Geos into the 1930s. One issue from the 1940s contained color plates of military insignia. I traded that one magazine to a junk dealer for a coffee can of unsorted "junk" coins. Not only did I extend my coin collection, but one of the coins turned out to be a token worth $20. I used this to buy an unabridged dictionary which I had been lusting after for a long time.

As a sophomore, I obtained a large box of old books from the late 1800s. Many of the books bore the personal library mark of Arthur J. Anders. When my roommate Arthur J. Anders IV came home I said, "Look, A.J.! This guy has the same name as you! Some relative?"

A.J. looked through the books and *went nuts!* The books were from the personal library of his great-grandfather, founder of the family fortune. Great-grandpappy's donation of books had started the early college library. Now the precious books were being ignominiously discarded like a pot-smoking frosh-stroking prof. The shame!

A.J. agreed. He trembled with upper class miff.

Well, not only did I sell the books to A.J. Anders Mark IV at a good price, but Arthur John and his numeral appeared in a number of pissed-off letters to the administration. (I helped him come up with a few good words like "ignominiously" and showed him how to make the remarks inflammatory.) Arthur John Mark III (A.J.'s father) threatened to curtail his substantial donations to the college. A sniveling apology was obtained. And I spent a lovely weekend at the Anders' family summer cabin, eating substantial amounts of uppercrust food, driving the speed boat and talking about business with A.J.'s father.

"You should write a book about these unusual skills and sell it in the specialty market," he suggested.

"Yeah, maybe," I agreed. "Uh...Yes, perhaps I *shall.*"

At the end of the weekend I went home with some lovely odds and ends. Try as I might, A.J.'s 17-year-old sister was not among my souvenirs. So, folks, WATCH THOSE LIBRARIES! Look for opportunities to make unique transactions. Swapping is fun, and *tax free!* What the government doesn't know won't hurt YOU!

So drop that Sumerian Pottery Philosophy course... dumpster depths are the true path to higher learning!

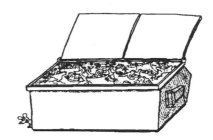

Chapter 9
Converting Trash to Cash

It's a great feeling to have somebody hand you CASH MONEY for an item you found in the trash. I rarely become as excited about my paycheck as I become over a few bucks from dumpster loot. It's like I *created* that handful of cash.

Actually, in an economic sense, I *did* create the wealth. Those items were set aside for destruction, effectively removed from the economy. By recapturing that wealth and injecting it back into the economy, I have *created* wealth. Best of all, when I receive cash it is *untaxed*. The underground economy triumphs again. Being "junk" or "secondhand," the items are sold for less than merchandise at the retail outlet. This contributes to driving *all* prices down, and allows people on the economic edge to acquire otherwise unobtainable goods.

Best of all, these goods allow people on the fringe of society to continue leading unconventional lives. When I sell an old pair of gym shoes for 50¢, some mega-corporation *suffers*. The tax-collecting retail whores *suffer*. The coercive tax-collecting apparatus *suffers*. And, perhaps, some radical has shoes to carry on his good works. I hope and will it to be so.

Many times I have loaded up my vehicle with clothes, toys, books, tools, etc., and looked at the stuff thoughtfully. I try to picture people using the stuff for a good purpose. Perhaps I'll pick up an item and say, under my breath, "God grant that the man who wears this hard-hat works for cash and avoids taxes."

This is another reason I donate unsaleable items to my local Goodwill. I'm not interested in charity for pathetic parasites. I see organizations of this nature as *radically subversive*. Think about it: as long as anybody can obtain clothes for a buck and books for a dime, they can avoid working a dehumanizing slave job to obtain their basic needs. Thus, they pay less taxes. The government pushes harder to collect more taxes, radicalizing *more* people. Far be it from me to provoke government coercion, even by a letter to the editor or a nod of consent. However, when I read that the government has committed yet another coercive act, I nod and say, "Good! Good!" Things do not *change* until a general level of discontent is achieved.

In our overly-regulated and coercive society, the majority of businesses are simply fronts for the government. When you shop at K-Mart, for example, you are supporting a multi-billion dollar entity that jumps through hoops for the government. If the government demanded all persons buying books show proper ID, K-Mart would slavishly obey the edict. Don't pity the "poor businessman." *He's a whore for the government.* You may as well be shopping at the IRS Store, Inc., instead of IGA, KFC, L.L. Bean, 7-11, *whatever.* The only "pure" capitalist is the guy selling stuff from the back of a van.

And so it is quite understandable that I take more pride in a small dumpster loot transaction than what's left of a paycheck. When I obtain that money *it's all mine.* And what's more, I have committed a subversive act.

But enough about the philosophical joys of liberated economic acts. Let's look at how to do these things.

My focus, of course, is on dumpster goods and I will address that in detail. However, I recommend reading *How to Make Cash Money Selling at Swap Meets, Flea Markets, Etc.,* by Jordan Cooper (published by Loompanics Unlimited). This book addresses the sale of all types of used goods in wonderful detail. You might also read *The Garage Sale Book* by Jeff Groberman and Colin Yardley (published by Prima Publishing).

Two general rules will help you profit immensely from dumpster goods.

1. Never, never tell people you obtained the stuff from a dumpster.

2. Avoid middlemen whenever possible. Sell or barter directly with the "end-user."

Rule 1 is tried and true, my friend. Telling somebody at a yard sale that you picked up these nice clothes "in the trash" is like casually mentioning that a smallpox victim died in 'em. People who saw a "great deal" a moment ago will drop the stuff like it burned their hands. Even if somebody looks right at you and says, "Aw, you picked this stuff up when somebody threw it away, didn't ya?" *say nothing of the kind.* Shrug. Roll your eyes. Pretend you didn't hear

the remark. Turn around and say, "Mind you kids don't drop that." NEVER admit the stuff was in a dumpster.

Here are some good lines you can use when pressed for a response.

- Junk I acquired when I helped a friend to move.

- Stuff I've had in the basement/attic-/garage/dungeon *forever.*

- Picked it up here and there.

- Grew on a tree. (For pesky people and/or wooden objects) .

Most people, of course, don't give a damn. They might ask something like "How old is this?" or "Does this still work?" You can either tell them you don't know or make up elaborate stories.

The *only* times I have been pressed for the source of items were at consignment stores. I used to sweat over this until I figured something out: their only concern was whether the stuff was "hot." Oh, a few times I have been questioned for no good reason by nosy, unprofessional consignment store clerks. However, in every case these were stores run by charities. Charity-based consignment stores are hopeless. More on this shortly.

Rule number two is less immediate but certainly more important. You will *almost always* do better to sell the stuff yourself or barter it. Junk buyers of every sort are cheap, crafty bastards. (The worst of the worst are *token buyers.* NEVER sell a token unless they show it to you in a book or a listing of some sort.) Extracting bucks from these junk-dealing bastards is like pulling away meat from a pit bull. However, a trade with these folks can be the highlight of your week.

A Barter Network

Bartering most types of non-food items is easy. Explaining how you happened across 200 t.v. dinners is very hard. Even if you find an enlightened individual willing to accept dumpster food, you risk revealing your sources. I'd rather use perfectly good food for *compost* than create a stampede to my favorite dumpster. In my own neighborhood I often grab items like old clothing that I will end up giving away. I do it to discourage the competition.

Mostly, I use dumpster food for myself and family. On the farm this allowed us to sell more homegrown produce. Another method that works is the "mysterious source." This is tough to pull off when you dumpster dive a lot. People are bound to see you. Usually your source needs some basis in fact. When Slash worked at the grocery store we made money hand over fist! We sold or bartered items like mad, explaining that our source had acquired the stuff. We would never say it was Jed, but our neighbors assumed it was. In a typical week we acquired more wealth in this manner than Jed's weekly paycheck. (This was the only reason he kept the job, by the way.) Our neighbors kept their mouths shut, too. They didn't want a stampede to *their* cheap food source.

The best approach to trade "suspicious" items is to develop a network far away from your source of supply. My father developed such a network while I was still a baby and Bekka wasn't even born yet. He regularly traveled to the VA hospital in a nearby city, where he struck up a relationship with the "Matthews," owners of an "antique" shop. Dad showed them a truck full of expired food items, and came home with a truck full of furniture, clothes and toys for the kids. Plus a few bucks for gas.

It was a scene repeated many, many times in years to come. I've seen little kids run into their houses shouting, "The Hoffmans are here! Hurry, hurry!" It's a real status thing to miraculously produce hundreds of dollars worth of "gray market" goods. A casual air helps increase the dramatic effect. Practice announcing this with a nonchalant air: "Say, I've got about a hundred frozen dinners and some crates of grapes. Want to make a trade?"

The dubious legality of these deals only adds to your stature. The funny thing is this: you *become* that mysterious, connected individual. Your announcement of goods to trade causes all kinds of hidden wealth to appear. People in printing plants offer books *that aren't even in stores, yet.* Pizzeria workers offer you half a dozen extra-large pepperoni pan pizzas. Poachers produce venison, still warm. Clothes, tools, job offers and skills are yours for the asking. Next thing you know your whole life has changed for the better. All you need is that "commodity connection" found in dumpsters to wheel and deal.

Years of trading and thousands of dollars worth of untaxed goods and services flowed between the Hoffman and Matthews families. Matthews helped us build an addition to our house and, years later, arranged for me to attend a far-right political camp in South Dakota. I shot real automatic weapons, hiked in the badlands, played games and met my first real girlfriend. Vehicles were fixed, families fed and clothed, pets and livestock obtained, medical care acquired, connections and skills traded.

Here's the kicker: A few years down the road Mr. Matthews asked my Dad if he would be interested in some food items that had been discarded. And next thing we knew, we were dealing t.v. dinners in "Hoffmanville" that had been dumpster dived in "St. Helga."

Cash for Trash

Remember, *avoid* the middleman. This means selling stuff yourself. It's not as hard as it sounds. Jordan Cooper's book provides much information. Here are some simple ways of selling goods, which I will cover only briefly.

Rummage Sale

Accumulate enough stuff until you can have a decent sale or profitably rent a space at a flea market. Remember, sell *everything*, even old shoes. You can even take magazines and sell 'em by the bundle or for a dime apiece. Somebody will buy almost anything. I've seen people buy used underwear with holes. (Of course, they paid in pennies.) If you can't have your own sale, go in with a friend or make a deal to use his garage. Distinguish your goods from his with cheap tape labels. Wheel and deal.

Postings

For one or two nice items, post a sign on a bulletin board. Churches, laundromats, apartment mail rooms and grocery stores are good locations. And you can sell almost any kind of furniture to college kids. They will pay five bucks for chairs with the stuffing dragging on the ground. Describe the item and always add a glowing term like "good condition!" or "like new!" A justification for selling it is good, too, like, "Moving — must sell!" Utilize bright colors and neat lettering. A Xerox of a Polaroid or the actual picture works well, too.

Friends

Sell items to friends, colleagues and co-workers as they express (or you anticipate) their needs. My mother knew the tastes, current clothing sizes and fashion likes and dislikes for every kid in her barter network — by heart! People would ask her, "Vernie, can you please get a pair of size ten shoes for Billy? He has a gym class and we can't afford any new shoes." And, incredibly, Mom could produce results. Of course, she also had a sideline hauling away rummage sale remnants. People knew she sold the stuff, but they would call her because she hauled the stuff away for free. But she would only haul away good stuff and demanded money if she showed up and the stuff looked like hell. (Personally — I would only haul stuff away for money.) We had a "permanent garage sale" in our old converted grainery, and it was a rare week when we didn't make at least $20 bucks. We usually made more.

Don't form a habit of giving things away to potential trading partners. I've seen shiftless people like the Ruben clan shell out cash money once they figured out the Hoffmans weren't going to hand out free slacks — school pageant or none. Once you give something away making a cash deal in the future becomes next to impossible. Many people believe we live in a "cashless

society" but they aren't talking about credit cards. I'll pack up stuff and give it to the Goodwill before I will spoil relations with potential trading partners. (I think nations should do this, too.) When people ask me for something they offer something to trade in the same breath. Only trusted friends can ask me for things on the basis of a future favor.

Like dumpster diving, barter is an art. Telling you how to trade is like explaining how to paint pictures. Not only is every deal unique unto itself, but you learn by *doing*. Find yourself some barter buddies and soon you'll discover a hidden talent in yourself. It isn't so hard. Strike up a conversation about bartering. Brag a little about having goods to trade. When somebody says, "Gee, I could use a few boxes of produce," simply say, "Well... got anything to trade?"

Many people become idiots at this point. Try to avoid talking to idiots in the first place. But a few people make a surprising and welcome response like, "I can introduce you to Kathy So-and-so." This is the art part. Determining the cash value of discarded doughnuts is easy compared to figuring out the "exchange rate" to convert expired cheese into an oil change. A glowing recommendation to an employer equals how many gallons of apple butter? See what I mean?

Pick trading partners with something to trade or extra cash to spend. Hit 'em on payday when the money is burning a hole in their pocket. There are certain people who will express a high level of interest in your goods, they will be charming and talk a lot but they have no cash and nothing to trade. I'd rather deal with a mean old cuss who has money or goods to trade. I once made a deal to have someone type a resume moments after I made an appointment to meet the same person in an alley behind our place of employment. I gave him a bloody nose and then told him to drop the resume by my house. Hey, business is business. At least nations have learned *this* lesson.

Pick people with a basic barter philosophy... people who don't expect favors on the basis of charity or their pathetic needs. Survivalists, many third world people (especially Vietnamese and people from Hong Kong), backwoods Mormons and many LIBERTARIANS are good bets.

Middlemen

As I stated previously, you will always do better to sell stuff yourself and avoid the middleman. I warned you. However, there are a limited number of circumstances when it is expedient to use a consignment store, flea market dealer or junkman. But BEWARE.

Consignment Stores
The way a consignment store works is simple in theory. You bring in the goods and the store owner sells the stuff for you. You obtain a cut and the dealer gets a cut. Everybody is happy.

But there's many a slip. For example, some stores demand 30% of your take. But that store may charge high prices and turn stuff over at a good rate. Another store may only take 25%, but the store is poorly managed and goods turn over slowly. High or low percentage is not necessarily an indication of good or bad management. Most stores will take a smaller percentage on a higher-priced item. For example, 40% on knick-knacks but 10% on high-quality furniture.

You determine the prices together — in theory. I'd rather find a dealer who knows his stuff and tell him, "What do you think? You've priced all my other stuff pretty good." But it's sheer torture to deal with a crazed old coot who determines a price too high and lets your stuff accumulate dust. Then he blames you for bringing him "crap." Or he sets the price too low and your good quality items sell for a song. You're left with 60% of a song. And about two thirds of such dealers will screw you in this manner. Worse, they lie about what they sold the item for and give you 60% of the fake price.

Most consignment stores have a 30 day limit, 60 for higher priced items. After that, they steadily drop the price. Most stores have some legal means of discarding unclaimed junk that doesn't sell. And, to be a consignor, you must

usually produce a goodly amount of stuff. If you have *that* much stuff, you could probably have a yard sale anyway. However, establish an account and you can often drop off one or two items as you acquire those things, even if your total stuff in the store drops to a small amount.

Consignment stores are great in one regard. Somebody else does most of the work. You can unload that stuff and keep your Saturdays for yourself. I used this "lazy alternative" when I was trying to accumulate a wad of cash by working many hours a week. But you lose control, and you *pay* for that convenience. I've seen stuff "disappear" on some fellow consignors. And most of these stores are in old tinderbox buildings.

On the other hand, having your stuff on display for 30 days (versus a one-day yard sale) can actually obtain you *more* cash, even with the dealer percentage. Consignment stores work well for busy people who can barely spare time to dumpster dive, let alone sell the stuff. It works well for small items you can't sell with a posting.

Always gather information and impressions of the store before doing business there. Understand the consignment contract fully, and inquire about how the stuff at the store is insured in case of fire, flood or theft.

But don't assume the contract will be applied to the letter. A consignment store may establish the *right* to discard property after 60 days, but many won't do it. Others may state they will not call you, even when you have some money from a sale. But, in fact, they might call you or send a postcard. A lot of stuff is put in the contract to protect the store from bastards. But don't assume they *won't* apply it, either.

Don't bring in stuff the store doesn't sell. For example, many stores sell *high quality* clothing like nice coats, but refuse to deal in old shirts, pants, etc., even if the clothing is in good shape. Some deal *mostly* with clothing, others mostly furniture, while the majority are "antiques and collectibles" oriented. You may find many "antique" stores are really consignment stores.

Clean your stuff before consigning. Most store owners have a pet peeve about this. Dirt marks in flower pots are O.K., a few stray crumbs in the toaster, but clean your stuff as much as you reasonably can. You'll obtain more money this way, too. If something came in a box or container, bring it. Bring warranties, instructions, etc. Confession time: I've often used appliance boxes obtained in the trash for my old appliances. There are exceptions to the cleaning rule. Don't try to strip old varnish off antique furniture and don't remove the dross from old silver. You can actually *reduce* the value.

Grab any "baby things" you see, even old clothing. Some shops specialize in this stuff. Baby clothes are grossly overpriced, always in demand, a good bet all the way 'round. (Better to sell them yourself or barter.)

Develop a good relationship with your consignment store based on mutual respect. If the dealer says, "This is no good," DO NOT become emotional. I have yet to hear any dealer in "junktiques" tell me, "Why, these things are wonderful! This stuff is worth a bundle! Where, oh, where did you find this?" The best response I've obtained is a raised eyebrow or a quiet statement that, "This may bring a good price." And many dealers, frankly, are masters of psychological browbeating. If you find yourself asking, "Why, oh, why did I think this stuff was worth anything? Maybe I should just sell it for a few bucks!" *you're being manipulated.*

Most dealers will attempt to purchase stuff by the lot. In an uncompetitive area, finding a consignment store willing to fill out the paperwork for a small lot is tough. And getting a good price is tough, too. Ten bucks and twenty bucks are some kind of "glass ceiling" for most dealers. Storming off in a huff or negotiating vigorously WON'T raise the price. These dealers (most of 'em) are mean-spirited or dumb. They know stuff takes a while to sell. Furthermore, they know most people with a trunkful of junk will accept their low offer.

You are much better off looking through the store (or around the flea market) for an item you like and then arranging a barter. Have the item in mind *before* showing your wares, and make sure it is an item the dealer owns, not a consignment. Express interest in the item but wrinkle your nose at the price. Then offer your items. Negotiate in cash and then ask about your items of interest. Play fair but play hard.

Last week a flea market dealer offered me $3 for a birdcage. Instead, I expressed interest in an old pistol holster, a coin and a book. We made the trade and I obtained stuff that would have cost me $18. Furthermore, it was *exactly* what I wanted. Best of all it cost me *nothing.* I may as well have found that holster, coin and book in the trash. I sat for a while looking at that lovely old coin and said, "Mine, mine, all *mine."* This sort of thing makes you feel so *clever,* so self-sufficient.

WARNING: Stay the heck away from consignment stores operated by charity organizations. These people are idiots. Worse, they have no profit motive. They might turn you and your stuff away because you are young and well-nourished. Meanwhile, some seedy old hag is given special consideration because she needs the money. Dropping prices or giving you a phone call requires a damned board of directors meeting. If you ask one of the employees a question they will invariably say, "I don't know. I'm a volunteer."

Used Bookstores

My personal favorite. I was tempted to put this first. I have already explained how to deal with coverless paperbacks. Now let's examine undamaged paperbacks and hardcover books.

YOU ARE BETTER OFF TRADING. Ask for cash and you will receive a pittance. (Unless the books are "rare.") Offer to trade and the whole store is your oyster.

Many dealers will demand 5¢ or 10¢ per book traded, or they will impose special rules. One dealer I know won't trade westerns or science fiction unless you bring some. Another has certain authors that can only be purchased for cash. These dealers are not mean or dumb, they are simply responding to market conditions. In fact, most of these dealers are very friendly people and brilliant conversationalists. They can steer you toward many a good book.

Some of these dealers sell magazines by the bundle, but that doesn't mean old copies of *Family Circle.* Many kinds of "soft porn" are very salable, especially old copies of *Playboy* and *Penthouse.* (Watch out. Some of these mags are worth a lot and that dealer won't tell you.) *National Geographics* are worth a bit, but tend to accumulate rapidly. If you find a collection of magazines in sequential order, or stuff like *Arizona Highways,* you can strike a good deal. Don't be surprised if the dealer wants to trade hardbacks for hardbacks, mags for mags, softcover for softcover. This is pretty standard. Tell dealers about your special areas of interest and they will set stuff aside for you, even call you at home. (Be specific so you don't waste your time or theirs.)

Many times I'll spot a title that I want, but I refuse to pay cash. So I leave with my other trades and remember the title. When I find some books in the dumpster I think to myself, "Hooray! I just found that edition of *From Here to Eternity.*" And, by golly, that's the title I obtain. Then I hug it to my chest and feel clever. I haven't paid retail prices for a book in years. Standard procedure in used bookstores is *half the cover price unless otherwise marked.* So if the book was printed ten years ago, you pay half the price the book sold for ten years ago. Why would you ever pay retail prices?

Scrap Dealers

Personally, I'm not willing to dig through a dumpster for individual cola cans... not now, anyway. But, of course, I have income from other sources. As a young person I spent many a day obtaining cash in this manner. I know an 11-year-old child who gathers up discarded soda cans so her mother can send her to beauty pageants. Whole families in Mexico support themselves in this manner, selling aluminum,

cardboard and other materials. If I were in need of cash in the worst way I would not hesitate to gather up soft drink cans.

Let me tell you, there is a *lot* of aluminum out there. All this publicity about recycling isn't affecting the majority of people. The average residential dumpster contains 50¢ worth of aluminum, *minimum.* Often, you'll find discarded stepladders, lawn furniture, screen doors and other aluminum scrap. Naturally, I won't turn my nose up at something like that!

You'll find plenty of copper in the form of cords on discarded appliances, old jumper cables, decorative items, etc.

Cut cords off discarded appliances and sell to scrap copper dealers.

Iron and steel scraps? *Forget it.* At $20 a ton this stuff is only a good deal if somebody pays you to haul it away. Glass? *Don't bother.* At $35 a ton you must collect 200 55-gallon drums of cullet to make a ton. And it all better be the same color. This isn't worth the effort even if somebody *does* pay you to haul it off. The stuff is made out of *sand.* Old tires? Hey, make yourself a swing, make planters out of 'em, but don't expect to be paid for your efforts. Paper? Only $10 a ton for mixed scrap. $35 a ton for newspaper. $150 a ton for high quality paper. Anybody discarding that much high quality paper would have to be in the printing biz — and they recycle everything at big printing places.

As a kid, I obtained the entire archives for a failed local paper. The stuff wasn't ancient, so I loaded up the truck and sold it to "Boa Brothers Recycling, Inc." I obtained 80¢. That's it. I threw the money in a soda machine, bought myself two colas and fumed.

Only aluminum, copper, and, in a few select areas, cardboard is worth the effort. Some cities have machines which allow you to insert soda cans and receive cash. These things are, in my experience, notorious cheats. The key to obtaining cash for scrap is to have somebody pay you to haul it away. I highly recommend *How to Earn $15 to $50 an Hour and More with a Pickup Truck or Van*, by Don Lilly (published by Darian Books). It seems to me that, by combining dumpster diving with the principles in Lilly's book, you can change that title to *How to Earn $30 to $80 and More, Etc.*

Other Outlets

A variety of specialized businesses exist which may be interested in your dumpster loot. Per pound clothing buyers, for example. Appliance and small engine repair places. Used bike shops. Army surplus stores. Campus book buyers. Typewriter repair shops. Coin, stamp, comic, baseball card and other hobby dealers. Some grassroots organizations may purchase office supplies. A pawn shop may give you a decent price, but don't count on it.

I know one guy who sells almost nothing but hubcaps. I know another guy and his brother who do nothing but purchase used batteries. (Sell used batteries for $2 to $5 apiece to auto parts supply dealers. Some junkyards buy 'em, too.) You might also find an auctioneer to sell your lot.

Many unusual outlets exist. Seek and you will find. My brother sold a broken, damned near useless pistol for $50 to a police "buy back" program aimed at reducing the number of handguns. Another time we obtained and immediately traded some discarded needles and

syringes to a "clean needles for junkies" program. We used the new needles to stock our survivalist medicine chest. Opportunities are unlimited. Dumpster loot provides a commodity to grease the economic wheels and start deals rolling.

Know the value of things. Whenever possible, snap up old antique guides, price lists for collectibles, books *about* collectibles, even old catalogues. These books, even when outdated, can help you determine which items are common versus rare. Whether you sell direct or use middlemen, read up on these subjects. With the exception of your next meal, what could be more interesting than the products of human society?

Note differences in price between retail and used. The minute you take a purchase out of the store, the price you can reasonably demand for that item drops by 30% to 50%. I believe this is due mostly to hidden taxes in production and sales costs.

There's another reason secondhand dealers of all kinds charge much less than retail: they *must* do this to attract a market. Frankly, most people would rather go to K-Mart and charge a leather jacket on their credit card than skulk around yard sales and secondhand shops. If you go in a retail store you know the item will most likely be there. In fact, there will be a selection. So you are paying somebody to do your thinking for you, to locate the suppliers for you and display it for your convenience. In effect, you're paying for a convenient source of supply... which is fine, if you can afford to pay 30-50% more than the actual street value. Do yourself a favor: become an expert on yard sales, flea markets, secondhand and junk shops. Take the money you save and send baby to medical school.

The Oddest Deals

The kind of deals you can make are as unique as YOU, as your individual circumstances. Some of the most artistic barter deals involve one-of-a-kind, impossible-to-duplicate circumstances. Only YOU can pull 'em off, because they are unique to you.

Take, for example, the time Jed and I managed to ruin a political career with some discarded x-rays. It wasn't really *barter*, but let me tell you about it.

Often I find things in the garbage that are unusual, personal, "neato" stuff that seems valuable but you wonder how, exactly. So it was when Jed and I found the x-rays. This was the dive where we acquired those used syringes. It was easy — all we had to do was locate a red, plastic "sharps" container. Let me tell you, we removed those needles with old salad tongs and had gauze over our noses and mouths. I highly recommend avoiding hospital waste except in extraordinary circumstances. We probably could have acquired needles and syringes through veterinary sources. However, Slash and I were all set to visit friends in the city and saw an opportunity to acquire these supplies for nothing. Heaven only knows what we did to the precious statistics of that "clean needle" organization.

Anyway, I saw an oversized folio in the hospital dumpster and picked it up. It was full of x-rays. The very first one was a skull with broken teeth.

"Check it out!" I said to Jed, my voice muffled from the mask.

"Neat!" he said.

So I dragged the folio home. The x-rays belonged to all sorts of people with lots of cool injuries. I thought these x-rays may have been used for instructional purposes, given the variety of fun fractures. The prize shot was a hand with an extended middle finger, with no apparent bone injury.

"Fun and games in the radiology lab!" I said to Jed.

"F--u--u--ck y--o--o--u!" he gasped in a "skeleton" voice.

Jed thought the x-rays were so much fun that I finally traded him the whole bunch for some fa-

vor or another. He decorated the window in his room and put the rest aside. It complemented his road sign motif fabulously, I thought. Jed even took Polaroid shots and sent the pics to some kind of "beautiful house" publication. He received a terse letter thanking him and stating that the photos and text were being sent to the "proper department."

"Psycho file," I told Jed.

"I just want to EXPRESS myself!" he replied in his "effeminate artist" voice.

Well, one day I came home from fishing and noticed a sporty little ragtop in the front yard. I tossed the stringer of fish in the sink and asked Mom who owned the car.

"Denise Bulltwaddle," my mom replied. (Not her real name, of course.)

"Daughter of COMMISSIONER Bulltwaddle?" I asked.

Commissioner Bulltwaddle was a notorious liberty-trampling bastard. Our neighbors were fighting him to the death over some kind of obscure zoning issue involving the proximity of livestock to a residential structure. An *abandoned* residential structure, I should point out. Only Bulltwaddle had a problem. There wasn't even a citizen complaint involved.

On several nights Slash and I kidnapped his garbage, looking for something incriminating to help our besieged neighbors. No dice. Lots of pizza boxes. We were set to try again.

"John," my mom said, "Denise is a nice young lady and you have to take her for what she is. And Jedediah likes her, so don't go spoiling it."

I wondered how *this* strange pair of bedfellows developed. Did the young Slashmaster find some sexy underthings and search out the owner? Was he pumping Denise for information? Was he pumping her for some other reason? Were there photos involved? Video? *Why was he holding out on his brother?*

Slash had a sign on his bedroom door which read, "DANGER! RADIATION! DO NOT GO BEYOND THIS POINT WITHOUT AUTHORIZED SAFETY EQUIPMENT." Another prize from the night at the hospital.

Slash answered the door and I had my first look at Denise. She had a punk haircut, black lipstick and tight leather clothing. She had skull earrings, a skull pendant, skull rings and little skulls on her t-shirt. A glow-in-the-dark plastic skull dangled from her miniskirt belt above torn fishnet stockings.

"Hey, John!" Slash said. "You met Denise?"

"Call me 'Daughter of Death'," she cooed.

"Charmed," I said, shaking her gloved hand.

So we all sat down to chat and the subject turned, naturally, to her Nazi pig Fascist father. Looking at her jewelry, a thought formed in my mind. I wasn't planning the demise of a county commissioner's career. I was simply seeking a way to make Denise happy and, perhaps, assist in the formation of a helpful alliance.

"You like Jed's room?" I asked. "The decor, I mean. *Death and Injury on the Road of American Dreams*, Jed calls it."

"It's O.K.," she said, nodding. "That magazine should have published Slash's photos."

"Stuck-up *pigs!*" I agreed.

"I really like the skulls," she said.

"You do?" Jed said, leaping up. I got a bunch more x-rays here in my dresser."

"Got any more *skulls?*" she asked.

I left quietly, smiling. And so the daughter of Commissioner Bulltwaddle left Jed's bedroom with a bunch of broken bones and teeth. Jed even parted with the prized x-ray we called "Bony Bird." And Denise a.k.a. Daughter of

Death promptly covered every inch of her large bedroom window... which faced the street.

When Commissioner Bulltwaddle arrived home from a hard day of harassing citizens he saw the bizarre window and *exploded*. He was, after all, chairman of the county hospital committee. Decorating his home with confidential medical records did not look good — and in an election year! How could his daughter do this to him? He charged into her room and began ripping down the x-rays, even breaking the bedside lamp. Denise had used the lampshade to display, for maximum effect, "Bony Bird."

Naturally, Denise jumped up to defend her property. And naturally Commissioner Bulltwaddle punched her in the face. Denise ran out of the house and straight to the county social services department. The social workers quietly transferred Denise to an overnight foster home. She called Jed, in tears. Jed promptly telephoned our neighbors. Our neighbors promptly called Commissioner Bulltwaddle and threatened to drag his name through the dirt if he didn't BACK OFF on the zoning issue.

And, amazingly, it worked. The very next day Slash and I sorted through twelve bags of courthouse waste and two bags of Bulltwaddle crap. We managed to obtain some social service papers which basically gave us clear title to Bulltwaddle's political soul. We also found, oddly enough, the x-rays. This is the only time we have dumpster dived the same items *twice*.

And we all lived happily ever after. Except Bulltwaddle, of course, who dropped out of politics and returned to the funeral parlor biz. Our neighbors presented us with several baby swine, in gratitude. And Denise presented Jed with a strange token of affection. The social workers had transported Denise to the hospital for x-rays, just to make sure Bulltwaddle hadn't busted her head. Somehow, Denise managed to obtain her own x-rays, which she presented to Jed. Eventually, she and Jed broke up in a disagreement about the morality of hunting animals, but they still lived happily ever after.

Obviously, it would be impossible to duplicate a situation of this nature. But it illustrates odd, creative use of dumpster loot... which YOU can do just as easily and just as well as I. Furthermore, it illustrates the value of dumpster dived information... which we will cover more later.

Remember, sell or barter the stuff yourself whenever possible. Barter requires considerable creativity but can be more profitable than selling for cash to middlemen. BE CAREFUL with the middlemen you elect to use.

Good luck... and good diving!

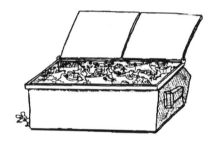

TWISTED IMAGE by Ace Backwords ©1993

MASTERING ADVANCED DUMPSTER DIVING TECHNIQUE — Chapter 437 "Avoiding Common Mistakes" | figure a WRONG | figure b WRONG | figure c WRONG | figure d WRONG | figure f CORRECT!

Chapter 10
More Dumpster Doings

In this chapter, I will discuss even more places to dumpster dive. We'll also discuss more aspects of self-sufficiency using dumpster goods.

The Manufacturing Sector

As much as I might like, I haven't had the opportunity to dive dumpsters from every type of business. Like you, I'm still feeling my way, so to speak.

The retail sector is more profitable to divers than the manufacturing sector. Why? Retailers depend on receiving goods and selling at jacked-up prices. To make more money, they find better ways to price or sell. They don't make much effort to "recapture" wasted materials, but try to order what they need when they need it. And their goods are *finished.* However, the people who produce goods deal with raw materials. They fret about waste and costs. You know what's in their dumpsters? Cigarette butts, coffee cups and empty bottles of Pepto Bismol. Dirty paper towels from the men's room. And paperwork, lots of obscure, dull documents.

But surprises lurk out there. I know of a business that makes specialized machines for packaging goods. A worker at this business used to bring our family big boxes of, say, individually wrapped cheese slices. They would run "product" through the machines to test the packaging functions. Most of the time this product was discarded, but the night shift always managed to make off with quite a bit. Our "contact" would trade us stuff like this for venison, seamstress service and various dumpster goods.

Obviously, this was an obscure kind of business. You probably don't have such a business in your area. But you may have something else just as good — if not better! The best way to find these goods is to get out there and hit dumpsters in a random manner to see what you might find. But logic and info-gathering can play a role, too. Think about the manufacturing sector in your town. What raw products do they use? What do they make and under what circumstances would they discard a "bad" lot? How would they "recapture" waste to cut costs? Would they *do* that? Is it cost effective?

I once went behind a poultry processing plant, hoping to obtain some feathers. (These add nitrogen to soil.) I was surprised to find out the feathers are processed and used to add protein to, among other things, chickenfeed and dog-food. Yech! Double yech! Watch 'em put this stuff in hotdogs a few years from now.

When you find yourself saying, "Why, they *can't* be recycling all their waste byproduct!" then you should go look and see. But don't be surprised if they are. A lot of food processors, especially, sell their waste to the people who make animal feed. I hope this doesn't "trickle down" to bakeries and grocery stores.

Another way to find out these things is to ask. This isn't so hard. Strike up a conversation with an employee. Ask him about his job and the manufacturing process. You can learn a lot this way even if you don't obtain a "dumpster lead." The employee who would bring us the cheese and stuff switched jobs, so we started diving the dumpsters at the plant. We wouldn't have known about this little plant if it weren't for that employee. Unless, of course, we had learned of it paging through the phone book. This is a good way to learn about obscure, hot diving spots. You may have some fierce competition from the people who live across the road from the plant. They may consider that dumpster their little secret.

Some surprisingly good junk can be found in the office waste from the plant. Businesses wishing to capture a market will bombard the "big cheese" at these factories with all kinds of samples — everything from plastic doodads to industrial lubricants to management texts. (By filling out postage-paid cards in obscure trade journals you can have these samples mailed directly to you!) Sometimes the big wigs run out and buy a sample of something just to examine it, discuss it with the other stuffed shirts, and discard it. You can also find golf shoes, attaché cases, postcards from Europe, all the bric-a-brac of the executive lifestyle.

Furniture Stores
Which Feature "Trade-Ins"

When you "trade in your old couch or dinette set" you may *think* you're getting a deal. Bull! They could hand out coupons just as easily. It would be a lot less work, too.

Once, some fairly well-off friends of my mother traded in a simulated leather recliner for a new chair. A few days later, they were driving through the alley and noticed their "trade in" tossed out amid a dozen other trade ins. They were outraged. For one thing, the chair had sentimental value. It hurt "Mr. Tillman" to see his Superbowl viewing companion tossed out like a piece of garbage. For another thing, they had paid a young man to bring the chair to the furniture place. (And let me tell you, I was glad to take their money.) Now their "great deal" didn't seem so great, after all. The "trade in" was a scam.

My mom assured the Tillmans that such a nice old chair could find a happy home, after all. We went to the furniture store, loaded up those discards, and brought 'em home. We kept the simulated leather recliner and used it for several years. The other furniture we sold for a few bucks apiece from our converted grainery. Whenever we saw ads telling people to trade in their old furniture we would go out of our way to dive that store. And, actually, they weren't discarding ALL the trade-ins. We figure 20% ended up in the alley. But that 20% meant quite a few bucks to us.

Here's a tip: Don't ever pick up an old mattress unless you need one. These items are virtually useless, and can rarely be sold for so much as a buck — even on the border. And you'll need a foam rubber pad to protect yourself from that one spring.

After Charity Sales — Help Yourself

I love a good church bazaar or a fundraising booksale on a Saturday afternoon. Of course, it's a pain in the ass to deal with all the volunteer salespeople. They always ask, "May I help you?"

and then never have answers. The point of their so-called "sale" never seems to be fundraising but rather the establishment of little committees and sale rules. But there are plenty of bargains if you can get at 'em.

Anyway, years ago somebody figured out these sales were good fundraisers. Many people will actually contribute costly items, believing the money will go to a "good cause." (Sure! If you call mass mailings, expensive office furniture and fat cat salaries a good cause.) But what happens AFTER the sale?

I'll tell you what DOESN'T happen. The Gray Ladies League doesn't call up every contributor to have them pick up their unsold items. They don't put the stuff in storage for next year's sale. After the Gray Ladies loot the leftovers somebody young and strong chucks everything in the trash. And, if it rained that afternoon or the sale wasn't well-publicized, you can find a LOT of decent stuff. I once found a box containing dozens of hard rock albums. Obviously, the albums didn't appeal to the housewives and little old ladies who were the primary customers at the sale. Or, perhaps, these "evil" records were removed prior to the church sale.

Another time, at a library fundraiser, I acquired a whole truckload of books. This was at the end of my senior year of college, and since I was leaving the area, anyway, I fired off a letter to the editor about this wasteful practice. What's really sad is that people donate items for a "good cause," often stuff with sentiment attached. They sacrifice the opportunity to have their own sale. And the human parasites who run non-profit organizations discard all this good stuff because they can't run a sale effectively.

If somebody in your neighborhood has a yard sale, don't forget to check their trash. You might find a lot of neat odds and ends. And, believe it or not, stuff can be found behind the Goodwill and Salvation Army. If you need a bunch of clothes for grease rags, go here. People often drop off nice pieces of furniture behind the Goodwill, and these donations tempt me severely.

Hotels — Soap City!

NEVER buy soap. You can obtain all the little-bitty soaps you need behind hotels. You don't need to settle for the used ones with pubic hair, either. If these little soaps are dripped on or steamed too hard in the shower, the maids toss 'em out, unopened.

Hotels which feature a complimentary newspaper are a good place to obtain your own subscription if you live nearby. This can be a very pleasant aspect of your morning walk.

Commercial Festivals, Grand Openings, Etc.

When I was a little kid, small towns throughout Minnesota would feature "Crazy Days" in late August. Merchants would drag out odds and ends of merchandise, put on silly costumes and have a big street festival. Unfortunately, as the festival matured the "odds and ends" mostly disappeared, leaving nothing but silly costumes and petty discounts. Still, wide-scale commercial activity often means lots of good discards. The only problem is digging through the wax paper cola cups, popcorn boxes and half-eaten hotdogs. This is particularly true of "grand openings."

Often, the days BEFORE a festivity are more profitable than the dive afterwards. It's not that more good stuff is discarded — just that it's easier to find.

Watch for "seasonal peaks" among certain businesses. For example, Slash wanted one of those grotesque rubber masks to compliment his road sign and x-ray room motif. We determined that it would be best to dive a novelty store immediately before and after Halloween. So we did, adding it to our "route." Halloween came and went, and still no rubber mask. But we found a lot of other fun stuff, so we kept hitting the dumpster. On November 16 (Jed's birthday), we pulled up to check the dumpster, by this time having forgotten about the goal of obtaining an expensive rubber mask. Slash casually opened the dumpster and shined his light inside.

"SHIT!" he hollered, and jumped back so hard he hit the side of the truck.

The lid crashed down, making a sound you could hear for miles.

"What's the problem?" I asked, jumping out of the truck with my bag blade in hand.

Jed caught his breath, grinned sheepishly, and pointed at the dumpster.

I raised the lid carefully, expecting a rat or some really gross dead animal with its eyes open. A delicious chill went through me when I saw the "zombie" mask staring back at me. I noted a rip that made the two eye openings into one large hole. Somebody had probably yanked it too hard while trying it on. I threw the mask at Jed's feet.

"Happy birthday, Slash," I said. "Or should I call you 'Crash'?"

And we found two more damaged masks in that dumpster. So, be mindful of heightened commercial activity, since it often translates into increased breakage and/or discards. But dumpster diving, like fishing, is more of an art than a science. People don't always discard stuff when and where you expect. I *frequently* find fake Christmas trees in the middle of July.

Your Own Personal Fire Sale, Earthquakes, Etc.

Natural and man-made disasters frequently mean discard-o-rama. Check the garbage cans and/or dumpsters after fire, flood, earthquake, riot, etc. Remember, these "disasters" can be quite limited and personal — such as one flooded basement or a little fire in somebody's kitchen. Insurance estimates take a few days or even weeks, so keep your eyes peeled for discards. The more affluent the business or residence, the more likely it will be insured and there will be a "discard delay factor." You are

also likely to have competition. People who wouldn't normally dive a dumpster are drawn to the spectacle of disaster, and they will poke through their neighbor's smoke damaged property with morbid delight. (Morbid delight that should be *yours!*)

Dumpster diving friends in California tell me that, after a good quake, you can pick up enough stuff to feed yourself for months... but, again, amateur competition is fierce. I've never seen this first hand and would love to hear from more divers in California.

The first time I took my wife dumpster diving, she located a slightly smoke damaged antique cola sign. We took the sign home and plugged it in, not expecting it to work. However, it *did* work and still works today — four years later. She has been offered as much as $50 for the sign, but keeps it for sentimental reasons.

When you dive disasters, don't forget about those interesting and oh-so-personal papers.

Construction And Destruction Sites

Dumpsters on construction sites are often fenced off. You may need permission to scavenge for building materials.

You can pick up all the building materials you'll ever need where buildings are being put up or torn down. Don't forget that an addition or a remodeling effort can also mean a lot of discarded materials. You can find a few of these items by keeping an eye on "do it yourself" stores — though construction sites are far more profitable. The only problem is that these sites are frequently fenced off and even guarded because of the tools, vehicles and supplies sitting around, not to mention the danger to adventurous kids. I find it's best to obtain permission from a foreman. Be respectful, say "sir" a lot, and ask for this person's permission to salvage materials. Be polite but persistent, smile, look 'em in the eye and NEVER act pathetic.

If the guy in charge says he can't have unauthorized persons on the site while work is taking place, ask if you can come back around 4:45 and grab a few materials while work is wrapping up for the day. Always indicate that you want "a few" or "some," then grab as much as you can. If pressed, however, look 'em in the eye and say you would like to fill up your truck.

Construction businesses *benefit* by your efforts, since they pay to have the materials hauled away. It sometimes helps to mention that, politely. Find out from the refuse company how much it costs and drop that figure, casually, like you're estimating. By the way, maybe YOU can do it cheaper and make a profit while keeping the good materials.

Don't be discouraged. Keep coming back, being extremely polite even in the face of rudeness or direct refusal. When one person grants permission, GET HIS NAME! This has saved my ass quite a few times when somebody comes running up, shouting, "Hey! Hey! What are you *doing?*" Be polite to this person, too, saying, "Sir, Mr. Blank indicated I could have some of this *waste!*" (Don't say "boards." Call it waste. Who wants waste?) Don't cop an attitude or you'll get your benefactor in trouble with his boss. Don't abuse your privileges and *stay away* from their other stuff. Thank your host. Once you have permission, scrounge daily. You might be able to sell some of this stuff, so grab all you can.

Destruction sites are good places to scrounge, too, and not as supervised. The people who do

this type of work are paid to take the building down within a certain period of time and haul it away. Some are paid as little as $1 and depend on the sale of salvaged materials to make a profit. They are after stuff like copper pipes, quality hardwoods and old fixtures. Most lumber is "scrap," and they will gladly let you supply "free labor" and haul it away — if you ask right. Bring a hardhat if you plan to ask permission to rip stuff out of the building. Offer to sign a "waiver of liability" if they express concerns you might sue them if injured. This is often their reason for refusal, even if they don't say so. Many times I have turned a "no" into a "yes" by remaining polite, persistent, and offering to sign a waiver of liability. Don't look clumsy. Don't use big words or they'll think you lack "common sense." Most people in this biz are self-made men lacking formal education, so approach them with this in mind. Follow their rules and grab, grab, grab. This is also a great way to get firewood.

Needless to say, it's easier to obtain access to good sites if you know somebody. Once you establish a good relationship you can scrounge again and again. Sometimes they will call YOU. The solid oak flooring in the Hoffman residence is a tribute to my Dad's ability to scrounge these sites effectively. (With the kids helping, of course. Again, never act pathetic, but a kid can open a lot of doors.)

There's an excellent book called *Building With Junk and Other Good Stuff, A Guide to Home Building and Remodeling Using Recycled Materials*. It's by Jim Broadstreet and is published by Loompanics Unlimited. This book not only addresses scrounging construction sites but details many types of unusual building projects.

Thinking Garbage

Growing up on a farm, we always had tons of materials around for different projects. Sometimes I would read about a self-sufficiency idea in a magazine, show it to Bekka or Jed, and soon we would be banging a project together. Everything we needed was in arm's reach or easily obtained. It wasn't until years later that I understood self-sufficiency projects are a hurdle for most people. If they want to build a chicken-coop, they start pricing lumber — as though the fowl will appreciate that "new house" smell and those expensive materials. The trick is to "think garbage." Translate the materials you need into "dumpsterese." When you read about a great project, ask yourself, "Where can I obtain these materials for free?"

My senior year of college, a friend made a wonderful dish of *pesto* for me and my room-mate, Scott. I knew right away that I wanted to have this dish all the time, and started thinking of a way to grow basil leaves on the balcony of our apartment-style dormitory. I eagerly told Scott about my idea, showing him some easy-to-make planters in a book.

"That's great," he said. "But where will we get all the *stuff?*"

Taken aback, I said I'd obtain the materials on my farm. But the tone of his question started me thinking. What if I didn't have a farm full of stockpiled supplies? What if all I had was this little apartment? Rather than grabbing the stuff from the Hoffman homestead, I decided to challenge myself and obtain all the materials right there in town.

I found some flowerpots discarded in the gar-bage cans at a cemetery. The wood was obtained behind a grocery store, from peach crates. I obtained the nails from these same crates. I found some discarded "Astroturf" behind the college stadium, to protect the wood balcony. The soil was dug up from a vacant lot, and baked in our oven to kill the bacteria and weed seeds. Thank goodness the utilities were in-cluded in the rent. I did cheat a bit by borrowing a hammer from the farm, but no big deal. And Scott "liberated" some sky blue paint with the help of a contact in the art department, trading her some dumpster dived hamburgers. She even painted a lovely white decorative design on the planters, watching Scott's skinny posterior whenever he walked by.

And we had our planters! Scott and I ceremo-niously redeemed some soda cans and bought the basil seeds. Yes, it was tougher than picking up all the materials on the farm. But we proved something. And we didn't have to call up our neighbors saying, "Got any lumber?"

So translate your projects into dumpster goods. It's good to have the stuff on hand, though, rather than running around looking for stuff. If you see some lumber, grab it. You'll think of a project later.

One easy-to-acquire item with survivalist applications is an old screen. You can use these to make drying racks and preserve your dump-ster dived produce. You can also dry seeds from, say, spoiled squash. Being hybrids, these seeds may produce many different kinds of produce — not all of these types are large or flavorful. But they are free, and good for neglected patches of ground.

No "Bag Lady" Decor

Of course, half the do-it-yourself ideas that I read about are pretty stupid. These are just ideas somebody thought up to make use of all the crap we can't recycle. Frankly, *anything* can be made into a lamp. That doesn't mean you need it in your home. Art is in the eye of the beholder, but I would rather not behold a wall hanging created from plastic cups.

Sick people dream up these projects, trying to save the earth and the precious landfill by tricking little kids into using piggy banks made from plastic bleach jugs. If somebody gave me a bleach-jug-piggybank, I'd cry and scream and throw a tantrum. I had a *real* china piggybank... with a chipped ear. My sister had a matching china piggybank with a chipped snout. Jed set-tled for a rubber "sad doggy" bank. All this stuff was dumpster dived, and we used bleach jugs and gallon cans to protect plant seedlings. Rainy day projects with the kids are fun, but why are so many "kid art" projects pure crap? I've seen many kids excited about a picture or painting; I have yet to see any kid truly excited because he made some "art" from a plastic beverage ring.

This is, quite simply, a highly advanced form of institutionalized child abuse. This sort of thing dulls a child's creative drive, makes it a pawn for some adult eco-game. Teachers who do this sort of thing should be flogged.

Back-To-Nature Snobbery

Digging through the refuse from a city is the last thing many self-sufficiency buffs want to do. And that's just what these people are — "buffs." Bark basket-weaving hobbyists. Survival is not a hobby — it's a goal. It's *the* goal, the *only* goal. If I knew there were some good things behind Sludge-o-chem, Inc., or a nuclear power plant, by god, I'd grab that item and use it to build a glow-in-the-dark rabbit hutch. I exaggerate to make a point.

The Amish, for example, strike me as ridiculous. If progress is so evil, why use a horse and buggy? Why not simplify to the point where people don't have domesticated animals? Let's *really* be righteous and take off all our clothes.

I look forward to the day when I can dumpster dive hydroponic solutions, supertrain public access cards, and books with holograms that are the right colors. Years ago I found the first hologram issue of *National Geographic,* and knew I had reached a dumpster diving milestone. Society will continue to advance, but we will still discard lots of great stuff. Hopefully, it will become easier and easier to achieve a comfortable level of survival using dumpster materials. One day, I believe, we will put a dumpster diver on the planet Mars. God speed the day.

Places To Avoid

Simple... none.

There are no dumpsters you should avoid unless, of course, the dumpster is locked, well-lighted, and rigged to explode. (We'll cover "obstacles" in the next chapter.)

Some dumpsters are more productive than others. But even a dumpster behind, say, a barber shop can have a "hot" day. (Besides, all that hair is full of nitrogen and minerals — makes great fertilizer. Six pounds of hair has the nitrogen of 100 pounds of fertilizer.) When my wife and I dive a dumpster, we hit the "hot" dumpster first. Say, for example, a bookstore behind a minimall. We hit the bookstore first, then one person drives slowly while the second person checks the Italian restaurant, beauty shop, and travel agency. (By the way — I hate that "perm" smell in beauty shop dumpsters. Eww! Give me kitty litter any day!)

A scenic vista for your viewing pleasure. When you raid "hot" dumpsters, raid all the other dumpsters in the area, too.

This is the kind of thing I mean by your "route." Say you have six "hot" dumpsters at points A through F. So you figure out the shortest route to dive them. But, along the way, you may encounter numerous other dumpsters. Dive 'em all. Go through alleys all the way. If you're young and nimble, run along while your partner drives. But don't dawdle. If you're old and slow you can burn up a lot of gas and energy for very small rewards. Whenever you are in a new area and have the opportunity, explore. You may find another "hot" dumpster. I like a nice mix of diving spots, for variety.

Of course, you should try to incorporate dumpster diving into your regular driving — no sense wasting a trip. But don't get so excited that you show up late for your job.

Keep that positive dumpster diving attitude. The good stuff is just around the bend and it won't cost you a thing!

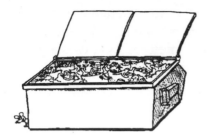

Chapter 11
Recycling Programs and Other Obstacles

This is a short chapter because, really, there aren't that many obstacles.

There are, however, some *enemies*. The biggest enemy is the ecologically unsound and eco-terrorist sponsored refuse compactor. In this chapter you will learn that "W.O.R.C. makes you free." That is to say, War On Refuse Compactors. You will also learn about the intensely retarded nature of municipal recycling programs, and why anti-dumpster diving laws aren't worth fretting about.

Your Enemy, The Compactor

Everybody's enemy, actually. The worst thing about compactors, from a dumpster diving point of view, is that perfectly good stuff is destroyed and/or rendered inaccessible. So let's explode some myths about the enemy.

MYTH #1. *Compactors save landfill space.*

Uh-huh. So, you think once that stuff is compressed it stays compressed like a lump of coal reduced to a diamond?

Your enemy — the trash compactor.

All the compactor does is keep dumpster divers from acquiring materials, like edible food, prevent the waste mongering business from requiring several huge dumpsters, which will mean a bigger refuse pick-up fee and an "unsightly" alley.

When that "compressed" refuse is unloaded at the landfill, it has just as much volume as the other refuse. Landfill savings are non-existent.

Think about this: the garbage truck compacts the stuff. The sanitary landfill runs bulldozers up and down all the refuse. So compacting at the discard point doesn't save any landfill space.

You know what saves landfill space? Ecologically committed divers like yourself.

MYTH #2. *Compactors are good for the environment.*

This wicked lie is merely an extension of Myth #1. Compactors are, in fact, bad for the environment in three ways.

- Dumpster divers are prevented from reducing waste bound for the landfill. This not only strains the landfill, but requires the growth, manufacture and distribution of more goods to replace these horribly squandered items.

- Compactors require electricity. 'Nuff said.

- And, perhaps worst of all, compactors increase toxic landfill seepage.

You know what seeps out of landfills and into the water table? A noxious, toxic, horrible mixture of pesticides, human waste, carcinogens, metals and other terrible stuff. The creature in *Alien* didn't leak juice this bad.

Any entity that every local governing body in the country calls "sanitary" is, obviously, an endless stream of filth. That's euphemism logic.

Allow me to explain this seepage thing. Visualize a half-empty can of insect spray. Somebody tosses it in the dumpster and the garbage truck picks it up. Maybe the can makes its way to the back of the garbage truck without exploding. Maybe, just maybe, the bulldozer at the landfill and the frequent garbage fires miss exploding the can. So the can is covered with dirt and garbage and corrodes ever-so-slowly, leaking that vile bug spray into the water table at a relatively slow rate.

By the way, all landfills leak to some extent, and the very worst of them in poorer areas simply spew. When the landfill management runs bulldozers up and down the waste, they *know* this will increase long-term seepage. But they don't have a lot of landfill space, so they do it anyway.

However, what if that can of bug spray goes in a compactor? Then, the can has several more opportunities to be crushed. The can arrives at the landfill, a mass of crumpled metal and liquid, and seeps into the ground water all at once. Feel thirsty? Good luck.

MYTH #3. *Compactors are private property and, as such, deserve to be respected.*

Compactors are the tools of eco-terrorists. And eco-terrorists deserve whatever they get. The only time you should hold back is when it's too risky. And, unfortunately, that is most of the time.

But, you know, things go wrong with machines and these things can be difficult to pin down. Some vagrants I know once disabled a new compactor by putting super glue all around the "on" button. Then they sniffed the glue and were arrested the next morning, passed out. Their hearts were in the right place, though. Compacting perfectly good food amounts to a policy of deliberate starvation of the homeless, not to mention the environmental impact. W.O.R.C. could be considered a form of "urban monkey wrenching." Hey, no decent person likes to wreck property. But there are times when we must. And, given sufficient justification to soothe one's conscience, it becomes good clean fun.

W.O.R.C. hard to keep the food free. Hard W.O.R.C. is its own reward.

Padlocks... Easy As Pie

First of all, not all padlocked dumpsters are secured all the time. It's a pain in the ass to run around looking for the key. And for what? To lock up the garbage? Who gives a damn?! Not somebody making four bucks an hour.

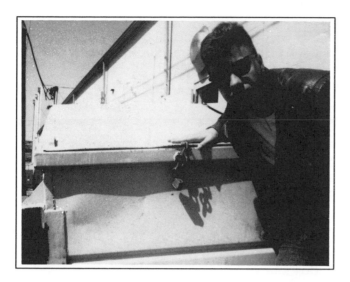

Even padlocked dumpsters are frequently left unsecured.

Most of the time the padlock is welded to a chain, so you'll have to gum up the lock mechanism to ruin the padlock. If it's not welded, cut that sucker off with a lock cutter. Dispose of it properly. After all, you're not an eco-terrorist like these bastards standing in the way of waste reduction. Slash thinks it's funny to replace loose padlocks with an identical padlock — which, of course, can't be opened with the original key. I say that's a waste of a perfectly good lock. But isn't humor worth a few bucks? This is especially funny a few hours before garbage pick-up.

There are a few rare dumpsters with built-in lock mechanisms. Give 'em blows with a rock. The business will blame the garbage men... hopefully.

Is all this stuff risky? You bet. Don't do it unless these locks and such are cutting off your food supply. Weigh the risks carefully. However, if you are a thrill-seeker I would point out to you that this sort of thing is more productive than, say, spray painting walls or throwing rocks at windows. *Fight the bastards by freeing up the food.* Before I moved to the southwest from the midwest, I went around my area and

snipped dozens of locks. Call it a community service.

Gates, Fences, Enclosures

Some dumpsters aren't accessible because they are surrounded by a fence. Other dumpsters are near truck bays, behind locked gates. A padlocked gate in front of a truck bay isn't there to protect the dumpster as much as the truck bay or loading dock. The business doesn't want somebody sneaking around the back of their property stealing equipment or breaking into the building. You may as well forget this dumpster unless faced with impending starvation.

This dumpster enclosure is attached to a building. You don't want to raid this type of dumpster because somebody will think you're a burglar, not a dumpster diver.

Fenced enclosures around dumpsters are a different matter. These are mostly found behind fast food places, and serve two purposes.

- Deny access to the dumpster so all those transients will come inside and buy the $2.99 triple decker. (Get real!)

- Hide unsightly dumpster areas.

Frequently, these enclosures contain equipment such as milk crates or grease encrusted stuff awaiting a good cleaning. Don't *assume* the enclosure is locked. Frequently, there is nothing but a simple latch. If it is locked, it may be more expedient to climb over rather than snipping the lock or gumming it up. Also, there's no sense snipping the lock and provoking the bun-stuffing bastards until you establish the value of that dumpster site. Besides, fast food people are robots. Cut their lock and they will replace it again and again.

A dumpster enclosure at a fast food restaurant. This enclosure has only a simple latch with no lock.

Another fast food enclosure. This one can be accessed with a little effort, but some are topped with barbed wire.

Check for a way under the fence first. Also, when you climb out you may be able to reposition some milk crates or other equipment to form a step for yourself. But be careful you don't slip, like Jed's friend, Teddy.

Jed was dumpster diving with Teddy, out of town, trying to find that hot burger bag. This was in St. Helga, where Dad befriended the Matthews family. Anyway, Teddy got inside the enclosure by climbing the fence, but the dumpster was almost empty. Teddy was trying to climb back out, standing on some milk crates, when he slipped and hurt his ankle.

Though in extreme pain, Teddy kept calm and pulled out his flashlight. He began flashing the light through a crack in the board to signal Jed, who was waiting across the parking lot in the truck. Jed climbed inside and assessed the situation. Sure, Jed could climb back out, but Teddy couldn't. And the enclosure entrance was padlocked from the outside.

"Take the tire iron and pry the latch off," Teddy suggested.

"I got a faster way," Jed said.

Slash wheeled the nearly-empty dumpster back a few yards, then slammed it into the door like a battering ram. He backed up and slammed it again, knocking the door off its hinges.

Then Jed, no *small* person, picked up Teddy and began carrying him to the truck.

"Lookit, Jed!" Teddy said, pointing.

Jed looked. The dumpster was rolling across the inclined parking lot, slowly picking up speed. It rolled out of the parking lot, bounced off a curb, and began to make its way down the darkened road toward Radison Lake public beach.

Jed followed in the truck to make sure the dumpster didn't hurt somebody. He heard a terrific "bang!" as he turned out of the parking lot.

The dumpster had made its way into the empty parking lot for Radison Beach, where it collided with, strangely enough, another dumpster. Both dumpsters rolled over a slight curb and ended up in a pile at the foot of a steep concrete embankment, in the sand. They landed in a position which, Jed told Mom, reminded him of two cattle mating.

Teddy and Jed looked for a moment, said "wo-o-o-ow!" and left to fix Teddy's ankle. To my knowledge, Jed and Teddy are the only two people alive to witness the mysterious dumpster mating ritual.

Retarded Recycling Efforts

Getting people to recycle is about as easy as getting people to poop in their pants.. and roughly equivalent. Dumpster divers will be in business for at least two centuries, even if recycling efforts go ahead full steam.

Let me explain. Long, long ago, happy little proto-humans lived in harmony with nature. They ate bananas and threw away the peels. When their stone tool broke they tossed it away. When they had to go doo-doo, they bent over the branches and cut loose. When a furry little proto-human died, they put his/her little hands over his/her chest, laid out some bananas and stone tools, and buried him in leaf refuse (so he wouldn't attract predators). Then they cried, ate a few bananas, went back to the trees and kept doing their thing.

All this banana-peel-tossing, stone-tool-discarding, cutting loose and proto-human-burying didn't hurt nature. In fact, the plants grew back more lushly than before where the proto-humans were buried, leading to proto-religious beliefs about the immortal nature of life.

But mankind progressed, discovering useful stuff like lead-based paint, styrofoam and insecticide. But he *still* insists on tossing his refuse around like a furry subhuman, expecting it to magically transform into a banana tree.

Something else happened, too. Germs were discovered. This simple but terribly important discovery caused a profound shake-up of human society. Before germ theory, spiritual forces were blamed for illness and disease. Certain objects were "taboo," or "unkosher," and certain people could, supposedly, administer

"magic" substances or perform ceremonies to keep these forces at bay. The all-pervasive nature of these beliefs cannot be over-emphasized.

But the germ theory changed everything. Suddenly (*too* suddenly!) people were told that natural forces were to blame for disease. Though remnants of the "spiritual forces" theory exist everywhere, the majority of people have accepted germ theory. However, they still *needed* certain objects or practices to be "unclean." So the germs became the focus of hate and fear formerly reserved for the dark spiritual forces. Garbage — once an emotionally neutral substance — has become the object of hatred and revulsion vastly out of proportion to its real danger. "Germs" have replaced the old "taboo" and forbidden objects and places. Germs have become the focus of irrational fear charged with spiritual themes.

Do you imagine that primitive man held his nose in revulsion when he smelled the droppings of an animal he was tracking? Of course not — odors were neither good nor bad, merely odors. Small children are fascinated by "poo-poo" and lots of bad-smelling things, until their parents teach them such things are repulsive and filled with unknown dangers.

Mankind must come to terms with the spiritual void created by the introduction of the germ theory. Garbage dumps are poorly-designed "forbidden zones" for "unclean" objects. We are being poisoned by our own festering landfills because we refuse to look refuse in the face, as though it had the "evil eye." And we are squandering the opportunity to recover millions of dollars worth of materials at the discard point — because we fear the dark. Because we fear poo-poo.

That is one reason I'm proud to be a dumpster diver. I believe the simple views I have just stated are important and real. Dumpster diving brings mankind closer to a rational society in harmony with the forces of nature and economics. We will never colonize the stars until we deal with our fear of "dirtyness."

And, besides, I'm just a kid at heart and I like to play with messy stuff.

Americans lead the world in waste. Our country is one big monkey house, only worse. Monkeys aren't afraid of banana peels after they toss them on the ground.

Some countries, like India, still use leaves to package their equivalent of "fast food." Many primitive countries rely on a variety of "organic" materials for packaging purposes. Weepy liberals point to these nations and say, "They're not advanced — they're backward! Boo, hoo, hoo." Of course, many of these "banana leaf republics" are acquiring a solid waste problem because they want First World products and technology. But it's a fact that America has the world's most shameful solid waste problem.

To deal with this problem, cities start "recycling" efforts. The word "recycling" has such a holistic ring — God forbid we should call it "resource recovery." Anyway, for a variety of reasons, *only manual separation of garbage at the point of origin leads to effective waste reduction.*

Garbage, you see, is complex stuff, a messy mixture of plastics, glass, metals, rubber, wood products, organic stuff, and so forth. It's hard enough to build a machine that will assemble an automobile when every part is the same every time and the end product is a high-value item. So "disassembling" garbage is tough. Garbage is different every time, and the end product of all this effort is worth only a small amount. If we could build machines smart enough to perform this complex task we would use these machines to perform surgery. Garbage would be — and *should* be — last on the agenda.

But Americans will move heaven and earth to avoid putting their hands in trash, even if they are radically committed to recycling efforts. What they usually attempt is the "big machine" approach. Simply build a facility which can take

a truckload of garbage and magically reduce it to neat little piles of aluminum, copper, plastic, paper, and so forth. Then they will sell that good stuff for... er... pennies on the pound. And the plant will pay for itself. Yeah, *that's* the ticket!

Crock Of Crap

What a crock of crap! This sort of thinking has no basis in reality. People are willing to build multi-million dollar facilities and keep 'em running on huge infusions of tax dollars just to avoid dealing with old banana peels. Every one of these facilities should have a large plaque on the front of the building with the words, "TO OUR FEAR OF POO-POO. ICK. YUCKY. BAD." There shouldn't be a groundbreaking ceremony for these buildings, but a ritual hand-slapping-and-loud-bawling ceremony.

But, weird as it sounds, the politicians who push these monstrosities on the public are, in a sense, right. These facilities are usually built after the local landfills are full and efforts to promote trash separation fail miserably. People hate and loathe their own waste products.

Oh, a number of people will gather up their newspapers, save their soda cans and such. A small number of weepy, pathetic individuals will wash individual jars and sort by color. But the vast majority will refuse to deal with their trash.

Working people say they don't have *time...* and it's true. Others say, "Why should *I* sort my trash when the businesses don't? When my neighbor doesn't?" And that's true, too. Apartment complexes say, "Why should we buy special containers? Let the city buy 'em!" People whine that they can't tell brown glass from clear glass, slick paper from newsprint, tin from aluminum. They hold up an example of, say, plastic and paper bonded together. "And what am I supposed to do with *this*?" they whine, self-righteously. God help me, I actually pity the politicians who are forced to listen to this monkey chatter.

And so, the people who wanted to make everybody sort their trash are tossed out, and some new bums promising creative solutions are tossed in office. Invariably, the "novel solution" they propose is to build a big waste processing plant.

"We'll heat the whole city with garbage!" they cry, like a crazed inventor with paint fumes on the brain. "We'll take all the recyclables and sell 'em! We'll build schools, hospitals... a Barnum Bulltwaddle Community Center!"

Pathetic. Really, we should study the long-term impact of landfill seepage on human brain cells. These shallow-thinking paper-pushers will invariably mention someplace like Finland, pointing out that "dem Finns" convert 60% of their waste to energy, heating their cities with clean-burning refuse. That's like pointing to a country like Bahrain and saying, "Why can't all *Americans* have a free college degree?" The answer to all of this starts with the words, "Because, stupid...!"

Many cities actually import European waste processing technology. Then, to make the alien facility run correctly, they hire hundreds of people to pick out the paint cans, coolant containers, aerosol spray cans, old motor oil in plastic soda bottles, and so forth. European trash is different than American trash. Period.

Mythical Markets

Furthermore, just because somebody, somewhere is buying paper pulp doesn't mean every community with a recycling program can readily sell these materials. Newspapers, in particular, are piling up in old airplane hangers, warehouses and other storage areas. Some cities are paying to get *rid* of their stockpiled recyclables. Plastic recycling, in particular, is a joke. Plenty of plastic containers now bear the "recycle!" symbol. As if! Hell, you can recycle DDT if you locate a market.

Worse yet are those ads showing "soil" produced from a treated disposable diaper. Yeah, right. In what obscure North Dakota settlement?

There's a name for this kind of bullshit: eco-porn. The recycling market *just isn't out there,* but businesses will keep pretending it is, they will keep sponsoring their Potemkin Village diaper recycling projects. People will feel less guilty about discarding their diapers and plastic jugs and the landfill situation will worsen.

Whenever I see one of those "organic disposable diaper" ads, I'm sorely tempted to go out, find a dirty diaper, and mail it to these people.

The Worst Part

The worst part is, surely, anti-dumpster diving laws and actual stifling of small recycling efforts.

Sooner or later, you see, the waste recovery plant doesn't make enough money. The people running the place whine, "We can't make money if people are picking all the aluminum and other good stuff out of the trash." Laws are quickly passed declaring garbage "municipal property" and dumpster diving is theft. Small recycling businesses are bought out by the city.

Think about the stupidity! Dumpster divers and small recyclers are working efficiently, recycling things and injecting money into the economy. The waste recovery plant lives off tax money like a junkie, sucking the local economy dry. Who gets blamed? The dumpster diver, of course. And when he stops picking through the trash, the facility *still* doesn't make money. And it will *never* make money because the whole idea is flawed from the start, based upon an irrational fear of garbage.

Laws Are Funny Things

Pass a weird law, and everyone is all excited for sixty to ninety days. Articles appear in the newspaper, praising or criticizing the new law. A token arrest is made. Everyone at the city council shakes each other's plump, hairy hand. Then the law is forgotten, except by anal retentive folks who write pissed off letters to the editor asking, "Why are people still diving dumpsters? Where are the police to stop this blatant infraction?"

If your city passes such a law, lay low for awhile. Then go back to business as usual. Seriously. You might try to obtain some damaging material on the sons of bitches who passed the troublesome law. But that's covered in the next chapter.

These laws suck, and I don't have answers about how to eliminate them. We can't rid ourselves of the drug and solicitation laws, despite vast numbers of people who would like to eliminate these laws. Fortunately, enforcing anti-dumpster diving laws usually ranks right up there with enforcing minor anti-smoking laws. When threatened by these laws, narrow your scope to your best dumpsters to reduce your risks.

And, of course, good luck.

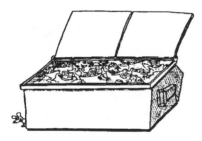

TWISTED IMAGE by Ace Backwords ©1993

Chapter 12
Information Diving

There's *so much* interesting material out there! If, for some reason, I were only allowed to salvage informative materials, I would still dumpster dive constantly. Imagine casually walking into the residence or business of your choice and perusing files, letters, photographs, *everything*. Imagine picking up a few "mementos" like a checkbook, credit card, cassette tapes, videos, floppy discs or you-name-it! Many times Jed and I have literally jumped up and down with delight, saying, "I can't believe it! This is *too* great!"

Thank goodness we had access to photocopiers at various jobs, because we were always mailing copies of documents to the local paper, as well as various gadfly citizens' groups. Some of the stuff was so hot we had to use gloves. Rubber gloves, that is.

In this chapter we're going to cover lots of good stuff: finding incriminating and/or damaging information about enemies, forms of ID, celebrity trash, dumpster dived coupons and other goodies for your amusement and enlightenment. And I mean YOU! "Information" is so broad, and the value of what you find depends so much upon your areas of interest, when you

take up "info diving," you'll be writing your OWN chapter — maybe a whole book.

However, let me preface the discussion of this delightful stuff with a word of admonition: dumpster diving is not the only way or even the best way to obtain information. Use your library — especially interlibrary loan. Peruse courthouse documents and newspaper archives. Use your local historical society and the Freedom of Information Act.

But dumpster diving is invaluable in two ways: (1) For finding nuggets of info that can't be obtained any other way. (2) "Serendipitous" information, or finding something wonderful and unexpected. That's a rush.

Trash-Picking Private Eyes

Most books I have read about private investigators mention trash as an excellent source of information. However, these books cover trash picking only briefly, with few anecdotes. I think private eyes are trash pickers more often than they admit.

Trash can tell you EVERYTHING about a person: What he eats. Where he shops. His income. His hobbies. His doctor and his medications. Where he works, plays and stays overnight. And, of course, you might get lucky and find something really good: A letter from his escaped felon brother. A note from a mistress. Old documents alluding to some long-buried scandal.

Dumpster investigations are also cheap. Following somebody around costs time and gas. If you simply grab the guy's garbage shortly before trash pick-up, you'll save oodles of effort. Hope your "mark" has his own garbage can. Sorting through the trash from an entire apartment complex is a bitch.

Personally, I try to avoid making enemies. Plenty of people interested in "survivalist" topics are vengeance freaks. They carry around a useless burden of misdirected rage, constantly taking offense at imagined slights. This attitude only encourages people to actually persecute the individual in question. I'm not a "vengeance freak," and so I've seldom found it necessary to investigate enemies. Furthermore, I'm not intent on defrauding anybody... though I can't resist trying to learn how. And the potential in dumpsters to investigate and defraud is vast.

I have accumulated dozens of checkbooks, bankbooks, credit cards, forms of ID, incriminating letters and photographs. This stuff is commonplace. You do not need to be at the mercy of your landlord, ex-wife, boss or slimy local politicos. And the information is cheap, cheap, cheap! Grab it and start analyzing.

Serendipitous Diving Spots

Photocopy Centers
Years ago, a photocopy center set up shop in Hoffmanville. I was thirteen, and had just completed a hundred page manuscript about an invasion from Venus. My father was trying to obtain a scholarship for me to attend a "gifted kids" program at the University of Minnesota. However, the scholarship form had to be filled out in triplicate, with photocopies of the kid's creative

efforts attached to each copy. At 5¢ a page, this was starting to be some serious change.

One day, my dad was making some photocopies relating to his VA benefits. The clerk handed him a little card, explaining that each time you made more than twenty-five copies the card was stamped. Ten stamps meant a hundred free copies.

My dad quickly grasped the dumpster diving implications of this concept. What happened to the stamped-up cards which were redeemed for free copies? Might a daring diver find some of these cards and redeem them again? He might!

It was late in the evening, and I was sitting in the truck eating a dumpster-dived jelly doughnut and drinking a cola. I was planning, in my head, part two of my *Venus Invasion* trilogy. I had peered in the dumpster for the copy center, but noted only bags of discarded paper. But when Dad walked outside and said, "Throw all those bags in the back. All of 'em," I just said, "Yes, sir."

We hauled the bags into the living room and began looking for the cards in question. Almost right away, we found half a dozen.

"Keep looking," Dad told me and Jed. "In a few months we may need more copies."

Jed and I nodded, and kept sorting. I saved some blank sheets for typing paper and tossed the rest in our blazing fireplace.

"What's that?" Dad asked, suddenly. "That thing with the star at the top."

I handed Dad a rather dark photocopy of a letter. Dad looked at the letter intently, reading through the excess toner. It was, as it turned out, a carefully-worded letter from the assistant city police chief to the chief of police. The assistant chief made reference to "selective enforcement" of traffic laws in favor of certain city officials. The assistant mentioned "grumbling" by the lower ranks of law enforcement and wondered if a tactful word to these lead-footed city officials might be appropriate.

"Well, shit," Dad muttered, and sat there for a while looking at the letter.

We didn't find any more "hot" documents, though we found some interesting and personal stuff that was fun to read as we tossed it in the fire. Anyway, Dad didn't do anything with the letter immediately. He figured somebody had photocopied the document for a reason — maybe this person was going to send a copy to the newspaper. So Dad watched the paper for a couple weeks. When nothing happened, Dad sent a copy of the letter to the local paper. And he managed to cause a small but intense controversy to erupt. It was a powerful feeling to watch the crap fly back and forth and know you caused all the trouble. Dad warned us in no uncertain terms to keep our mouths shut. As poachers, dumpster divers, and welfare form fiddlers, we had no problem knowing enough to remain discreet.

Some years later, Jed and I were doing the "route" by ourselves. I reminded Jed about the "ticket controversy," and wondered aloud if we might find something similar in the garbage from the copy center. So we kidnapped the garbage and perused it. Though we didn't find anything as hot as the police letter, we found enough interesting stuff that first night to convince us of the dumpsite's potential. After that, we grabbed the copy center garbage a few times a month. Call it a public service. Some people are public watchdogs. We're public trash hounds.

Eventually, we found stuff that was better than the police letter. We found lots of stuff that wasn't so scandalous, but still informative. Some of our information erupted in the papers; most didn't. But, hopefully, hard questions were raised, rumors started, and the entire weekends of certain politicians ruined. A good controversy would cause us to dive the dumpster several times a week, hoping for more "pay dirt." We'd sit in our converted grainery, feeding an old Franklin stove with the "reject" papers, formulat-

ing the wildest plots. These are some real fond memories for me.

Years of hitting these photocopy places has taught me a lot. First, these places are hot where info is concerned. Dig through papers from an office and you'll find mostly crap. However, a wide variety of people use photocopy centers for important papers. These people frequently make several attempts to adjust the copy size and level of toner, not to mention aligning the papers correctly. Often "paste-ups" of originals are discarded. Lots of times sensitive copies are ripped in half, and both halves are discarded. Easy as pie!

The sheer variety of these dumpsters is wonderful. You'll find everything from little old ladies complaining about dry cleaning to sensitive memos from the campaign headquarters of congressmen. One of these days I expect to find a ransom note. You can't expect much monetary gain from these places — but it's fun. Try it and you'll like it.

Photo Processing Places

I promised you pictures of people you know engaged in sexual acts, didn't I?

Photo processing places charge a lot for photos, but the number of perfectly good discarded prints is amazing. As they adjust the color and exposure of their print maker, thousands of prints are discarded. These aren't fuzzy, discolored photos, either. The photos are often stuck together, however. If they won't come apart easily, dunk them in barely warm (not hot!) water. Gently separate the pics and let them dry. You'll also find empty bottles of chemical crap and lots of metal film casings. You may find a soda can or two and discarded bags of fast food. Pray that nobody at your local photo processor consumes shelled sunflowers while running the machine. Or chews tobacco.

It's best to dive photo places in small towns. You'll obtain more pics of people you know. However, even in big cities people tend to use the nearest photo place. My college was in a fairly large city, but practically the whole campus used an "Hour Foto" down the street.

Photo developing businesses are a great spot for "info diving." Here, a dumpster diver displays photos of a sexy classmate in lace bra.

There's a number of things you can do with these photos, not all of which are malicious. Frankly, I like to have photos of people I know. Naturally, I can take pictures of these people. But dumpster diving is like having a person hand you pictures from his/her album, saying, "Here's a nice shot of me with my niece. Oh, here's an ugly shot of me in the hospital. Here's me in a string bikini. Here's me exposing one breast..."

Once, a girl dumped me because of my wicked ways, leaving me without so much as a wallet-sized pic. I saw her going into the Hour Foto one day, and kidnapped their trash that very night. I managed to obtain some great pictures of this girl, which soothed my broken heart a lot. A few times I have found photos of *myself* taken at parties. Finding a picture of yourself in a black plastic bag is always a jarring experience. Stare into dumpsters long enough and they stare into YOU.

Analyzing dumpster photos is fun. Documents are pretty straightforward but photos call for intuition and insight. Scott and I used to sit

in our dorm doing this, saying stuff like "Gee, she said her parents had money... but what a crappy kitchen!" Or, perhaps, "Who is this guy? I thought she went with Paul So-and-so. Hmmm." It's good clean fun — except when you find something really hot. The kind of people who disrobe in front of a camera don't care if the photo processor gets an eyeful. And this is more frequent than you might imagine, especially around colleges. Really, you must dumpster dive a college-oriented photo place right after Spring Break. You must. I insist.

Some places will develop the film but refuse to print the x-rated prints. However, thanks to the automated process, the photo place doesn't know which pics are x-rated until they print a few. And, frankly, most places don't give a damn. You might even find a clerk or two with his own "special album." I saw a case like that on a "tell all" show. The problem wasn't his album so much as the fact he dragged it out at parties.

People are surprisingly uninhibited in front of their own cameras. Ask Gary Hart. You might break a scandal wide open and make history.

Government Offices
People are compelled to give all sorts of information to the government. This information may be especially valuable in a small town setting. I've always found county seats to be a fascinating dive. Many times Slash and I have retrieved forms from the welfare office and obtained the financial details about our friends and neighbors. Invariably, we would find these people weren't declaring income from lucrative "odd jobs." Since they weren't our enemies, and since we were more dedicated to the principles of economic freedom than the rewards of snitchery, we would simply put these documents in big ol' manila envelopes and file 'em for future reference. We had boxes and boxes of odd papers labeled "MISC. DIRT."

One time, I recall, we came close to using this information. Jed was participating in a play after school hours, and one of the Ruben kids had to stay after class every day for some stupid infraction of the rules. "Billy" Ruben's big brother

"Elmo" would pick him up with the "Rubemobile," an old Cadillac held together with baling wire. Jed was a husky young lad, but Elmo was big as a baby whale. He would find Jed, waiting for his ride, and rough him up. Dad wouldn't let Jed carry his "bag blade" to school, so Jed was taking quite a beating. Dad broke his "non-interference" policy to call up Mr. Ruben and told him to instruct Elmo to keep his slimy flippers to himself.

"Aw, who gives a shit?" replied Mr. Ruben — liquored to the gills, as usual. "Kids will be kids. Your kid just can't take it."

"And your kid is no kid," Dad replied. "I don't call twenty-one a kid, even if he is just an eleventh grader."

"Fuck off, Hoffman!" Ruben replied.

"Listen up!" Dad said. "I didn't want to do this, but you're leaving me no choice. You know your gravel pit?"

"Yeah?" Ruben replied, sobering up a bit.

"You're selling hundreds of dollars in gravel to a certain construction company and not declaring it to the welfare people, *aren't you?*" Dad asked. "And you've been doing that for years."

"What makes you say that?" Ruben asked, quite alarmed.

"I know it for a fact," Dad said. "And I don't give a shit whether you cheat the government. More power to you. But if Elmo puts his hands on Jed one more time, I will turn you in faster than a bartender will throw a wooden nickel back at you. Try me."

Dad hung up. And Jed never had any more problems with Elmo Ruben.

Personally, I'm a "memo man." Blackmail is not my thing. I'm more interested in controversial interoffice memos that I can provide to the press for scandal purposes. And I'm very fair in one regard: ALL public officials are fair game.

So many times I have seen "grass roots" organizations or individual "gadflies" doing battle with city hall. Their usual approach is to pore through public records, looking for stuff missed by the press — $50,000 for conventions and office furniture, that kind of thing. This is an excellent approach. But anybody willing to spend hours and hours digging through public records should be willing to invest an hour or two perusing documents which were supposed to be *destroyed*. Yes, I mean the garbage.

Public records are on file for years or even decades and centuries. Garbage — a fragile, lovely thing — lasts but a moment. Personally, I would run straight for the trash and save the public records for a cold, rainy day.

One small problem, folks. Government buildings *almost always* have some kind of garbage security. When it comes time to spend the money in the "physical improvements" budget you can bet some brown-nose will suggest a big ol' fence around the dumpsters. These are the same bastards who can't do their dull little job without new equipment, new furniture, frequent "seminars" and so forth.

However, plenty of government offices are set up in temporary locations without dumpster security. Social security offices located in minimalls, for example. I've seen everything from the local branch of the FDA to small chambers of commerce located in these unsecured locations. There is so much good stuff, both in the form of fascinating information and in the form of delightfully blank forms, stationary, ID blanks, and lots of unused stamps on unmailed envelopes. These places are also a cheap source of such things as manila envelopes and file folders. Why pay retail? Why pay?

Police dumpsters are particularly fascinating — and risky. As a journalist I would go to the police station every few days to read the Initial Complaint Reports, or "ICRs." These were fascinating and fun, but rarely important enough to make the paper. After my job with the paper was finished, I would grab the police garbage every so often just to keep current. One of our friends was having a terrible time with a certain neighbor always reporting "suspicious" activity at his perfectly innocent parties. I found a partially-typed ICR and was able to confirm the identity of the complainer. Of course, this was just a document I found by accident. Most things you find are like that. Be flexible and creative.

Private Organizations

These places have less dumpster security and can be almost as much fun as government dumpsters. And, of course, many of these organizations deserve to be investigated and opposed. I know one woman whose son was severely mistreated by a drug rehab program. She dug through the organization's dumpster looking for incriminating documents and the names of people she could contact to see if their children reported similar mistreatment. She found both.

I saw another instance where this sort of activity made a tremendous impact and caused all kinds of problems for the organization in question. This involved Jed's friend, Teddy, and his mother "Mrs. Spooner."

Mrs. Spooner was a nice lady who made delicious homemade desserts. She was the pretty young widow of a man who didn't believe in insurance. She was a dedicated mother, a reasonably open-minded lady. There was only one issue that could transform her into a foaming-at-the-chops wild woman: the abortion issue. She said rosaries daily for the unborn children. She spent many hours a week working for her cause. So, when an organization doing abortion referrals set up shop in St. Helga, Mrs. Spooner was beside herself.

Previous to this, Teddy and Jed spent many happy hours rummaging through papers in our converted grainery, which they jokingly called "INTEL II." One day Teddy found a paper related to Mrs. Spooner's issue. It was only a photocopy of an article, but the paper started Mrs. Spooner thinking. What sort of documents might she find behind the "Women's Choice" office?

Poor Teddy really screwed himself. His radically dedicated mother insisted on raiding the office dumpsters nightly. She even dropped her second job and took up home daycare so she would be free to raid and rummage.

The first night they encountered a padlocked dumpster. This was no problem for Mrs. Spooner. She purchased twenty-five padlocks of that exact type and tried all the keys until one worked. Then she returned twenty-four of the locks for her money back. After all, this was her *cause*. Monetary risk was no object.

Mrs. Spooner found all kinds of good stuff. Receipts from contractors doing work in the office. (Promptly boycotted by members of Mrs. Spooner's church.) Names and addresses of women and girls requesting information and services. (Promptly contacted by members of Mrs. Spooner's church.) Names and addresses of individuals working in the office. (Promptly harassed by members of Mrs. Spooner's Church. And I mean *harassed*.) A copy of a letter to the editor concerning the harassment. (Mrs. Spooner called the paper, impersonated the writer, and asked that the letter not be published.) And so forth.

The people at the office were going nuts trying to figure out the source of these "leaks." They even accused each other of being "spies." One worker was fired and filed a lawsuit. The reputation of the organization plummeted because they couldn't keep anything confidential.

Mrs. Spooner never left a clue about her activities. She would carefully replace the garbage after rummaging. Eventually, the office closed up shop... much to Teddy's delight. Mrs. Spooner detailed her methods to other people involved in her cause. The info was, I presume, passed all over the country.

Mrs. Spooner's activities (love 'em or hate 'em) illustrate an important principal: the value of dumpster information is in the eye of the beholder. Where a person does his dry cleaning may be of no importance to me. However, to somebody intent on destroying that person, such information is like gold.

Jed went along with Mrs. Spooner and Teddy one night, just to learn about practical dumpster-based harassment techniques. Jed looked through the papers a bit and thought to himself, "This is shit." The only thing of value to Jed were the names of local girls who apparently went "all the way." But these papers were like gold to Mrs. Spooner.

Jed had even more bizarre dealings with Mrs. Spooner's group, dealings that involved a human fetus. But I'll tell you more about that later.

Thy Neighbor's Mail

The best place to obtain mail is in residential areas. But another interesting dive is the local post office.

Plenty of people with post office boxes open their mail on the spot, read it, and discard it. This is especially common with bills they do not intend to pay.

One of my friends was particularly fond of ordering books and magazines, then refusing to pay the bills. He would claim he never ordered the books, somebody else did it to harass him.

One day I was rummaging through some discarded mail and found a bill for more than $50 from a seller of "quality publications." The bill, which was unopened, bore the address of my friend.

Just for kicks, I called my friend and pretended to be a bill collector. At first, "Fingers" told me to take a leap. He had not ordered any books. Somebody was ordering stuff to harass him, and he didn't owe the company a dime or the return of their books.

But I was firm. I stated that we had matched handwriting samples at Fingers' bank, Third Digit National. I said we had matched the fingerprints on the mail-in cards with prints on file at the local sheriff's department. We were, I said, prepared to file a class-action suit on behalf of

one dozen other book companies. We were also considering charges of mail fraud. If Fingers didn't pay his bill TODAY, he could expect an arrest warrant. Other companies might just roll over, I said, but not "Quality Publications," makers of fine books and magazines.

I was so convincing that Fingers capitulated. He offered to mail me a check. I said a sales rep would come by and pick up the check personally.

Let me tell you, Fingers was pretty pissed off when I showed up on his doorstep and said, "So where's my fifty bucks?"

Of course, I find bills of this type all the time. But, like I said, fraud isn't my thing.

Another thing I like to do is "check" electric bills. A couple years ago I called the electric company and disputed my bill. The company said they would send out a meter reader to double check the reading.

"What does this cost me?" I asked.

"Nothing!" said the utility rep. "It's a free service."

The next day I saw the meter reader doing his thing. After that, I would have my meter rechecked monthly — just because I hate the electric company. But I also "check" about ten other people's bills every month. Call it a public service. One time I became so pissed off about somebody's bill that the company rep said, "Do you want me to just disconnect your service, Mr. Garcia?"

"Why...no!" I said, quickly.

I also like to call up the credit card companies and find out everyone's current balance. All you need is their account number (on the receipt) and you can hazard a guess at the zip code if they live in your area. Of course, to cancel the account or change the address all I would need is the cardholder's date of birth, social security number, a few things of that nature. A few weeks later I could ask that a card be issued in the name of my "girlfriend" and sent to the fictitious address.

Yes, it's a bit more complicated than I've described. But it's amazing what you can do with a little information. And, of course, Loompanics Unlimited provides many quality publications on these subjects. Dumpster diving can reduce your risk of detection while obtaining information. Read books on these topics and "translate" the contents to dumpsterese.

I'm more of a "performance artist" than somebody intent on fraud. For a couple of years I thought it was fun to call up a certain credit card company, request some harmless information about obtaining a credit increase (which they would sometimes offer me on the spot), then mention casually that I'd read certain terrible things about their company and didn't know if I wanted to stick with such an awful bunch of people. The abject groveling was always amusing.

Postage-paid return envelopes are also oodles of fun. I use these envelopes to distribute my favorite propaganda nation-wide. Several times, Slash and I have written amusing tracts detailing bizarre theories. Our favorite is "cartoon balloting," a theory we actually believe whereby individuals express political discontent by voting for cartoon characters. We drop the tracts in the envelopes, then sit back and laugh while huge corporations pay to receive our weird propaganda.

Back to the post office. Most of the discarded mail consists of "bulk" mailings and "stuffers." However, I often find nice stamps from other nations, even return envelopes with "good" stamps on 'em. Soak in warm water, peel off carefully, and dry. Use paste to attach the salvaged stamp to your own envelope.

Lots of different companies have "contests" in which they give away a small prize like a fax machine or a dozen raspberry bushes. Many of the companies are kind enough to provide a postage-paid return envelope. Personally, I'm

not a "contest nut." My mother and my wife are both contest nuts, and also frequent winners. So, whenever I see a discarded contest offer I grab it. Small scale contests (the "free fax machine" type) can be modified with a simple name change. For stuff like sweepstakes, my wife calls the company's "800" number or drops a postcard in the mail and asks to be put on their list. In the past she has won watches, dogfood, a phone, $50 and all kinds of other small freebies. She's convinced that, one day, she'll win something really big. I think this is all pretty silly, but I'm not one to talk since I'm a "junkmail junkie."

Another interesting item found in discarded mail is porno. There's more of it among your tightly-buttoned neighbors than you might think. Many of the companies which sell porno mail out their catalogs every month, seldom making any real changes. The customer soon becomes bored and discards the catalog as soon as he receives it. You can tell which envelopes contain porno catalogs by the following criteria:

- Moderately thick, big enough to hold a small catalog.

- A return address like "Entertainment Products" or "XYZ, Inc." Something that sounds fun, but cryptic.

- Outside markings like, "TO BE READ BY ADDRESSEE ONLY." This is a dead giveaway.

You can enjoy paging through the catalogs yourself or simply make a note that your neighbor, Mr. Blewnose, actually likes "red hot transvestites in high heels." You might even obtain a clue as to which of your neighbors likes to get wild and naked.

More rarely, you'll find the actual magazines. Lots of people receive gift subscriptions to magazines they don't enjoy. I know just where to look every month for a copy of *Penthouse* and *National Geographic*.

Dumpster dived return mail cards can result in nice freebies by mail, like this butter substitute and 5 pound boxes of mix.

Specialized Publications

This is a good place to mention this favorite item of mine. As I've stated before, I can't get enough of 'em. I've learned about "pain compliance techniques" reading *Police* magazine, I've learned how vendors position soda machines for maximum sales, and I've learned about new products designed to look "fresh and tasty" after eight hours on a "hot rack." Many times I've saved myself money and hassle by learning the "tricks of the trade." Everybody's trade, that is. Libraries sometimes have a few of these publications, but dumpsters contain stuff you can't obtain anywhere else with the same ease.

But there's another reason I love these magazines — free samples! I always send away the postage-paid return cards requesting "more information." Frequently, this "information" arrives in the form of a free sample. This is especially true of the food industry.

Ah, but you're saying, they won't send me anything because I'm not a chef or a hotel manager or anything. Not true! I've been doing this thing since the age of eleven, and it doesn't make any difference. Here are the main "tricks."

- Never fill out any of the "extra information." For example, most of these return mail cards

ask for the name of your business, number of employees, annual sales, etc. Skip all this stuff, even if the card says you *must* fill it out. Half the time they will send you the stuff, anyway. DO sign and date your request. But otherwise, provide ONLY your address.

- Never put your phone number on the card, even if they say you must. Half the time they will send you the stuff anyway.

Sometimes they will find out your number and call you. If they call you, immediately say, "I don't recall providing you with my home phone number." Be vague. Be rude even. They won't be able to tell you from their other "prospects."

Sometimes, I actually engage the salesman in conversation, learning from these enthusiastic people all the latest advances in "ultra-sonic vacuum leak detection." My little junk mail hobby keeps me abreast of high tech stuff so I can keep writing good science fiction.

Certain kinds of companies will always call. Insurance people. Various investment firms. Non-profit organizations. Military recruiters. Some places that offer to "send information" will do nothing but call you. They don't even have materials prepared for mailing.

I always say, "Look, I'll be happy to look over any information you might send me by mail, but I'm a busy guy and don't want to talk with you right now."

- Send away for everything. Your name will end up on everybody's list and you'll receive junk mail you never even requested. You'll have the opportunity to receive free subscriptions to some of these publications. Then you won't have to locate them in dumpsters.

- Ignore expiration dates on return mail cards unless the dates are several years old. "Expired" cards are frequently answered and, in any case, it costs you nothing.

Lots of people hate junk mail. Well, that's them. I *love* the stuff. I love the slick illustrations, the intense sales pitches, the paper I can use to start the fireplace. It gives me a sick thrill to set apple crates ablaze with a bonded paper letter that begins, "Dear Drilling Site Supervisor." In fact, I love all my unearned salutations.

I also have another reason for this hobby, a sort of weird and highly subjective reason for liking junk mail. It's my way of "taxing" corporations. Corporations cheat me by collecting sales tax on the items they sell. Traitors. Who hired them to do the government's work?

So I "tax" the bastards right back. By demanding information I cost them time, postage and materials. Sure, I can't extract as much value from the materials as it costs the corporation to mail the stuff. But they can't obtain much value from my tax dollars, either, since the money goes to the government. So it's a draw. Every year, businesses extract hundreds of dollars in taxes from me and I extract hundreds of dollars in postage from them. When they let up, *I'll* let up.

When the government taxes you, it obtains information it will probably use against you. That's what I do, too. I tax corporations and demand information I will probably use against them. I am a sovereign government unto myself. And it's grand. I recommend it highly.

Two more reasons I love junkmail are the "big brother" factor and "micro-immortality." Sit back and let me explain this weirdness.

First, regarding "big brother." Every time you make a move it goes in a file or computer. Somebody can come along and use this information against you. So, you can do two things. You can make as few moves as possible, covering your tracks, hoping information isn't being accumulated for use against you. Or you can mess with their minds by putting yourself on every list in sight. Apply for everything. Ask for information about everything. Put your name everywhere. If you want information from the Libertarians, ask for info from the Communists,

Expansionists, Grassroots Party, Demopublicans, Repocrats, *everybody*.

I pity the individual or computer that tries to accumulate information about me. I've requested info about every group, every piece of equipment, every profession, every school, every product, you-name-it. I've sent in warranty registration cards for hundreds of appliances I don't own. I have joined virtually every group that doesn't charge to become a member. My computer file must resemble... well, a pile of refuse.

Most of these things I do to obtain information. But there's the "big brother" factor, too. My philosophy about that is don't hide — mess with big brother's head.

The second reason is "micro-immortality." I like to put my name on mailing lists for the same reason people carve their names on a tree, have children, write books... I want to live forever.

When my paternal grandfather died, we had his mail forwarded to our farm. Five years later we were still receiving mail in his name. It was like a little piece of him was still alive. I like the idea of my name and address multiplying like a living thing and spreading through databanks forever.

So, you're wondering, how much mail do I receive? Well, I average twenty-five pieces a day. When I was a kid I was more dedicated, going to the library monthly and ripping all the "bingo" mail-in cards from dozens of magazines. Back then we left a bushel basket in front of our rural mailbox. I once received a hundred and twenty-seven pieces of mail *in one day*. People would call from New York offering to sell me gold futures. My mom always got a kick out of telling them I was an eleven-year-old kid who enjoyed receiving such mail. Our living room looked like an embassy office under siege as we burned papers in the fireplace.

Once I received a mailing from a company selling software that could, so they claimed, eliminate "unproductive" names from mailing lists. People, in other words, who just liked to receive mail... like me.

I sent the company a letter detailing my sordid life of junkmail addiction — right down to the part about digging in the trash. I pointed out that, obviously, their software had some holes in it if they sent an offer to somebody like me. For a fee, I wrote, I would show them how to pinpoint "unproductive" names. They never wrote back. I should have just mailed a copy of my letter to *Omni*.

I'm not above using my initials or calling myself "Joan Hoffman" in order to receive more mail. When I was first dating my wife, she told me one night that she had to run out and obtain a certain painkiller used mostly by females. I said, "Here. Somebody mailed me a sample."

You know what else you can find in discarded mail? Stickers! I'm crazy about stickers and stuff like Easter seals. In fact, Easter seals are the official stamp of my sovereign self-government. The letters always arrive "postage due," IF they arrive, but is it my fault nobody recognizes my government? Anyway, I love stickers. I love to put stickers on everything, creating whole walls of stickers. The overall effect is a sort of mad, sticky colorful commercial and political collage. I think it's a hoot to see a Pat Robertson bumper sticker next to a sticker that says, "YES! SEND MY FREE X-RATED VIDEO!"

I'm just a kid at heart who loves stickers. And discarded mail is full of stickers, even the kind that say, "THIS HOME PROTECTED BY BOND HOME SECURITY ALARM CO." My home is protected — shouldn't yours be, too?

Here's a tip: one of the best places to obtain mail without skulking around behind the post office is the "mailroom" of a local apartment.

Discarded mail played a role in an encounter I had with some rat poison. I was diving in a lovely residential apartment complex when I spotted a discarded box full of personal papers. The papers seemed to be in regard to local police matters. There were also dozens of copies of *Po-lice* magazine and a nice pile of books about self-defense subjects. Only one problem: the box was covered with a white, powdery substance. It wasn't flour and it wasn't cocaine.

I found an empty box in the same dumpster and began to remove the articles one by one. I shook off the majority of the rat poison and dropped the items in my box. When I arrived home I took the box to our bathroom and set it in the tub. I turned up the fans to keep the air moving and began to carefully wipe each item with a wet rag. The letters I simply read and tossed away, since they were interesting but not valuable or incriminating.

Our pet rabbit peeked in the bathroom doorway. She loves to chew on paper.

"Scat!" I said, closing the door to keep the dumb bunny out.

Well, perhaps I should have just locked the bunny in her cage. The air didn't flow as freely once I closed the door. And the materials were so interesting! I must have spent an hour in that bathroom wiping off books and perusing letters. I was wearing rubber gloves but probably inhaled small amounts of rat poison.

Suddenly, I felt dizzy. My heart was beating rapidly. I felt like vomiting and passing out at the same time. I left the bathroom and sat down.

Luckily, the symptoms passed in half an hour or so. Dying from rat poison in a lavatory would have been an inglorious end to a master diver. I had saved all the books, so I just took the rest of the "poison letters" and tossed 'em out.

Anyway, be warned. There have been many times I've boldly rummaged amid rat poison, grabbing one or two non-food items. But I try to limit my exposure. Those interesting letters just tempted me too badly.

Celebrity Trash

Ah, many is the time I've thought of moving to Hollywood, California to begin a life of beachcombing, screen writing and dumpster diving.

Ah, Hollywood. Good supermarkets... nice climate... and celebrity trash with a decent resale value. That's the life for me.

Most of my "celebrity" dives have been limited to the trash of local officials. However, I once worked in a luxury hotel and had the opportunity to scrounge numerous souvenirs from celebrity visitors. These included rock stars, movie stars and political candidates. I found a quick market in the "groupies" hanging around the hotel.

If you have a celebrity in your area, be aware of this: celebrities frequently have private security and these people are all-too-aware of the tendency of celebrity garbage to "walk off." Some years ago I watched one of those "real people" shows which featured a celebrity-trash raider. He had souvenirs from such notables as Richard Nixon. Unfortunately, he didn't sell the stuff — he just made artsy collage things.

I'm sure YOU are smarter than that.

Trashy Art

People who prefer pictures of Labrador retrievers and duck decoys to abstract expression have been calling modern art "trash" for decades. But some art has actually gone in a trashy direction, using discarded materials for the purpose of artistic expression. Some of this is just "bleach jug piggybank" stuff... in other words, a bleeding heart attempt to save the world from plastic beverage rings. But plenty of "trash" art is quite good. And it's good for this reason: you can find ANYTHING in the trash.

Your means of expression are not limited to eggshells and fish skeletons... despite the illustration I saw, recently, for a "trash art" show in a major southwestern gallery. How insulting, I thought. What a stereotype! Would they show a black artist painting pictures of fried chicken and watermelon? Dumpsters are so much more than food refuse. I've been diving for more than a decade and have yet to see a fish skeleton with

head, despite the cartoons which show this to illustrate "garbage" in visual shorthand.

Where else but a photo processing dumpster can you acquire thousands of color prints for use in art projects... FREE? For absolutely nothing, you can obtain wild stuff like telephone parts, giant cardboard cut-outs, uniforms, and all the glass, plastic, paper, cardboard and cloth you can haul off. FREE! Dumpster diving *artistes* don't have to work food service jobs to buy materials. Get your supplies FREE and spend your time doing what you love.

As you've noted throughout this book, members of my diving clan love the colors, smells, tastes, and textures of discarded materials. Often we use the items we find in a certain way to produce a unique look or feel. Art doesn't have to be in a museum to touch your soul, to stir something within you. I've been drug free my whole life, but many times I will see a weed-filled crack in a sidewalk or a discarded piece of clothing in a vacant lot; I stop, look for a while, and say "Wow" a lot.

When Slash decorated his room with road signs, x-rays and rubber masks... when we covered the side of a building with multi-colored license plates... when we used a brightly-colored shelving unit for a chicken roost... these were instinctive integrations of art into our daily lives. The art is highly subjective and may only appeal to us. But it appeals to us at a deep, wordless level. We love it, and we don't give a shit what the world calls "real" art.

Perhaps my happy descriptions of discarded commercial bric-a-brac in a hog farm context don't strike a chord with you. But maybe you become ecstatic over military items. Western things. Old phones. Orchids. Cacti. Van Gogh. Seashells. Whatever is a unique turn-on to you, well, THAT'S how I feel about the inside of a dumpster.

Jed and I know one guy who is crazy about antlers, bones and horns. Whenever we found a discarded rack of antlers we would sell 'em to this guy for some quick cash. He made wonderful objects for sale, but the point of these sales was to allow him time to work with his favorite things: antlers, bones and horns.

Personally, I'm crazy about fossils, soda cans, and old newspapers. I would love to have a huge home decorated with these objects. But that's me. Other people like antlers or old Chevys.

My aesthetic sense and dumpster diving matured together. I can look in a dumpster, see a circle of discarded cigarette butts smeared with lipstick, and say, "Wow!" Ancient Indian pottery, on the other hand, makes me say, "That's nice." If I weren't a writer, I would probably make art objects using dumpster materials. And I would be happy as a hog with a bucket of pasta. Ancient Indians probably felt the same way about their pottery.

Both Slash and I have pursued careers as writers, though Slash is very involved with what could be called drama or "performance art." Bekka has become a sketch artist and painter who also creates expensive quilts with modern materials. So I suppose all three of us grew up to become artists, sharing a common "dumpster eye."

Dumpster diving was never purely a matter of survival to us. We did it because *we liked doing it,* because it was incredibly fun and, weirdly enough, beautiful and mind-expanding. We did it to defy stale convention even as we hid our activities. To us, discarded tomatoes weren't just food, they were beautiful. We were eating (and smelling and tasting and seeing and touching) beauty. And garbage. And there was no line to distinguish beauty from garbage.

I believe that every expression of human beings is *art.* I see life and expression and soul (as well as despair, ignorance, death) in every aspect of human activity. And it turns me on. I happen to be a dumpster diver, and I do it with the mind of an *artiste.* The diving part is art... the barter... the careful storage of materials... the use and consumption of these items is wonderful and beautiful (as well as ugly and sad) and I *love*

it. When I find a can of chunk tuna, I don't merely eat it... I commune with its essence. I own its beauty and substance (and its pain and ugliness) FOREVER. I throw away the can with reverence.

Art and the use of discarded materials goes back a long way. For example, Edvard Munch painted "The Cry" on a discarded piece of cardboard. The artist who can only acquire discarded materials — perhaps because he has defied every aspect of the status quo — speaks to me more deeply than the artist who is coddled by rich patrons, his every need for materials instantly fulfilled. Such an artist sells little pieces of his soul for a morsel of meat.

Dumpster art is truly "art from the edge." And that's where art belongs — on the frontier of human experience.

Perhaps you're not into drawing, painting or making statues. But dumpster diving can sustain you while you pursue other creative goals. Those "other goals" are YOUR art. And, if dumpster diving helps you in your art, I will be glad, indeed.

The Future: Floppy Disk Diving

In the last few years, I have seen an amazing dumpster phenomenon. People are discarding floppy disks and computer related materials *by the ton.* Often, I'll grab a box full of "interesting papers" and find it's all "computer stuff."

Finding a floppy disk is like finding a whole file cabinet full of papers — but in a compact, easy-to-use format. I've accumulated dozens of these disks, looking forward to the purchase of a personal computer with the proceeds from this book. Often, I'll take a few discarded disks to work and use the office PC to check out their contents. I'm always careful to use our "anti-virus" program with this mysterious software. Once, I actually found the infamous "PLO" virus.

"No wonder they threw it away" I thought.

But most of the time all I find are hundreds of personal files, computer games and programs — all *kinds* of neat stuff. I'm saving these disks for use and in-depth perusal at a later date.

A group of "hackers" at my college who ran afoul of college authorities confessed that they had obtained "passwords" by rummaging through discarded papers from the Admin building. Yes, computers are here to stay but so is dumpster diving — and many times the twain shall meet.

Cassette Tapes, Videos, Records, Etc.

You can find lots of these materials discarded in residential areas. Personally, I won't purchase cassette tapes or videotapes when I know I can obtain 'em from the trash. Just erase the junk on the tapes and record your own stuff. But many times I find tapes or videos I want to keep. And, as CDs edge out the old vinyl records, you can find more and more old albums hitting the bins. Sometimes I find tapes with arcane conversations recorded — these are always fun.

X-rated videos are almost always broken. Either they are made cheaply or people do rough things to them.

More Neat Access and Freebies

A few chapters ago I mentioned a free oil change that I scammed thanks to dumpster diving. The trash is filled with such freebies for the sharp-eyed and daring diver.

Lots of businesses and organizations give "freebies" or "special deals" to certain customers. These aren't exactly like coupons, which are generally available. I mean stuff like "free car wash with your next five gallons! Hot wax, too!" Things like ski passes. Mail-in rebates. Stuff you can turn in for free photo processing, a nickel cola or a dollar sandwich. I've even found school lunch tickets. (There is a free lunch!!)

Sometimes considerable boldness is required to use these freebies. More often it's no big deal. For example, a local grocery store cashes payroll checks — for a fee. When you cash your check,

they charge $2. However, you are given a coupon good for $2 off a $25 purchase. Amazingly, people discard these coupons. I use 'em myself or sell 'em to neighbors for a buck. And the trash is full of this sort of thing, all of it unique to your area. But, again, a sharp eye and boldness is required. Don't ask dumb questions or act like something is "wrong" with your coupon, voucher, seat ticket, etc. Act like you're in something of a hurry. But do have an excuse ready in case you are vigorously questioned — which hardly ever happens. I frequently gain access to movie theaters, museums, food fairs, etc., by using things I've found in the trash. NEVER pay full price for anything if you can avoid it.

The best place to obtain such things is, of course, in the trash of the business in question. Our "free photocopies" scam is an excellent example. It also pays to read your "mailbox stuffers." When you find something free, pig out! Store up fat for the dry season.

Dumpster Coupons

It pays to clip and organize coupons. Plenty of "super dooper shopper" books exist that tell you how to use coupons to the max. And you've probably read an article or watched something on television which features a lady obtaining $200 worth of groceries for $1.27. Well, it is possible but, like everything else, requires some effort and some smarts. The principles in these "coupon books" can almost always be summed up as follows:

- Organize, organize, organize. Clip your coupons and organize according to category. Don't just save 'em all in a big cookie jar. People with a HUGE food budget will find it more effective to organize by category AND brand name.

- Sit down with your weekly shopping circulars and plan your grocery shopping as carefully as a *coup d'état*. Don't walk into the store without planning your every move. Don't buy on impulse. Don't shop while you're hungry. Know your local stores like the back of your hand.

- Find a store that "doubles" coupons. This is the "secret" to all "super" shopping. Double coupons and buy the item while it is on sale. Buy great deals in volume.

- Don't be conned. The whole point of coupons is to lure you into buying a certain product. Don't be suckered. Use coupons YOUR way, to YOUR advantage, and don't play the game the way Madison Avenue wants.

I should point out that coupons are a great thing to have while skulking "bargain" carts. A few times I have found stuff like, say, a slightly squashed box of cereal for a buck. If you have a 50¢ coupon, and the store doubles coupons, the item is YOURS for nothing but the tax.

So, you're wondering, where do you obtain coupons? Well, besides your local paper, coupon clubs, magazines, and freebie "bins" in the grocery stores, there are a number of dumpster sources.

- Residential garbage. Collect the coupon sections from discarded papers. If there is a 20¢ coupon for canned beans, and a store that doubles coupons is selling those beans for 50¢, it makes sense to obtain as many cans as you can for 10¢. So you need *lots of coupons*. So start looking for those discarded Sunday papers.

- Post office trash. Coupons should be considered a "bonus" in addition to all the interesting mail. I wouldn't dive a post office for just coupons, unless it was a little post office with lousy dumpster security.

- Apartment complex mailrooms. When I pick up my mail, I check out all the coupons and bulk mail deals. If something looks really good, like a one-per-customer-with-coupon grocery special, I grab a handful of discarded circulars. I'll go back again and again in different store locations and "stock up" on the really good specials.

- Recycling bins. The coupon circulars are discarded along with old newspapers. Since many people subscribe to papers from distant cities, you can frequently obtain coupons not generally available in your neck of the woods. This is true of apartments, too. You're also doing the recycler a favor by pulling all that "slick" paper out of the newsprint.

This is also a great way to obtain a newspaper "subscription." A paper that is read twice is truly recycled.

One out of ten coupons is redeemed, which means nine out of ten end up in the trash. If you are a die-hard "coupon clipper," trash is a goldmine!

Don't forget to check discarded magazines for coupons, especially women's magazines. Dumpster diving discarded mail is a good way to obtain "consumer surveys" which you can mail in to obtain free coupons. I receive dozens of special coupons in the mail because I constantly mail away these surveys. Naturally, I lie about my marital status, my pets, even my annual consumption of aspirin. Not only is this a chance to obtain coupons, but a cheap opportunity to confuse the hell out of "big brother."

Cigarette companies are in a class by themselves. Not only will they mail you all kinds of samples, but you can save their "proofs of purchase" for mail-in freebies. In their desperate attempt to create more nicotine addicts, these heavily subsidized companies are one of the last sources of true rebate freebies. Where can you find proofs of purchase? Everywhere people discard cigarette packs.

It's rare when you can use a "proof of purchase" from a carton instead of a pack. But I have seen such deals from time to time, and hope to see more. Empty cartons can be obtained by the dozen at convenience store dumpsters. That's about the only thing you can obtain at a convenience store, besides cardboard boxes and, rarely, magazines.

The Art of Mail-In Rebating
I'm good at this, but I'm a mere novice compared to Teddy's spitfire mother, Mrs. Spooner.

Being a devoted and determined single parent, Mrs. Spooner was a regular household whiz. She read all the housekeeping books, magazines and articles. Mrs. Spooner took great pride in her ability to make a good party dip or remove a spot. So it was no surprise she was a regular whiz at mail-in rebates. And roughly 75% of the "proofs of purchase" that she used came from the trash of households where she performed cleaning services.

Mrs. Spooner had two government surplus file cabinets from the WWI era. These cabinets were once dark green but Jed, Teddy and I decorated them with a layer of colorful stickers. It really added a "coupon" look to the files. Once we explained THAT to Mrs. Spooner, she calmed down considerably and even liked it.

Anyway, Mrs. Spooner's motto was "NEVER THROW AWAY A PROOF OF PURCHASE OR A GROCERY STORE RECEIPT." She might have added, "WHETHER IT'S YOURS OR BELONGS TO SOMEBODY ELSE." She had one file just for receipts and all the rest of her drawer space was devoted to proofs of purchase. She had an old cash register with a fat roll of paper and a new ink ribbon, ready to manufacture a receipt if the need ever arose. The majority of her drawer space was devoted to proofs of purchase. A small card index (salvaged from a fire-damaged library card catalog) helped Mrs. Spooner keep track of what was in the drawers.

You name it, Mrs. Spooner saved it. Soup and canned pasta labels. Boxtops. "Points" from cake mix boxes. Certain UPC codes with the words "proof of purchase." And so forth. When a toilet paper company demanded the face of their "tissue kid" as proof of purchase, she began saving those things. We pasted a "tissue kid" to the outside of that particular file, as an artistic statement.

The minute Mrs. Spooner obtained a mail-in rebate offer, she made a beeline for her drawers. And, nine times out of ten, she had enough proofs of purchase to mail in the rebate on the spot. She would mail one for herself at her home address, one for her job headquarters (using her nickname), and one for her elderly mother across town. If the offer wasn't limited to "one per household," she would take advantage of the mail-in as many times as she could.

Mrs. Spooner sort of learned this stuff by accident. When she was newly widowed with a little baby in a strange northern town, she would use mail-in rebates just to obtain mail. (Mrs. Spooner and I shared many a laugh and many a tear over our mutual addiction to junk mail.) She was willing to part with 50¢ or a dollar for "postage and handling" just to get a package in the mail and talk to the UPS man. So she wasted a few dollars obtaining "ketchup cookbooks" and "guides to using Spanky's Instant Taters" until she discovered the UPS man was a pothead. But at least she had the cookbooks and guides, which led her in the "art of housework" direction. When she wasn't working or looking for a better job she would bounce Teddy on her knee and dial "800" numbers *all day*, just to have an office worker in Atlanta (her hometown) tell her, "It's nice here. Sunny and 80 degrees!"

Mrs. Spooner's attempt to fill lonely hours had all sorts of unexpected benefits. She found out that, by dialing these free numbers, she could obtain free samples along with the requested "information." And the rebates began paying off in spades as she obtained free coupons for a gallon of milk or a pack of toilet tissue. Mrs. Spooner began evolving the moment she asked herself, "How could I do this *better* and get more of this good stuff?" Soon she purchased the old file cabinets at a weekend yard sale. Grabbing other people's trash just sort of happened. She saw a proof of purchase on a box of fishsticks and just couldn't control herself. It was, she said, like watching a dollar bill waving to her from the trash can. Next thing she knew, Mrs. Spooner was a die hard dumpster diver. Proofs of purchase leap at her from dumpsters the same way paperback books leap at me.

Mrs. Spooner used mail-in offers to obtain virtually all of Teddy's toys. Little Teddy probably never had a teddy bear — ironic as it may seem — but he had every stuffed cartoon character from every cereal commercial on television.

Usually companies demand "postage and handling." This may actually exceed the value of the "free" item, especially stuff like mugs or bowls. But a buck fifty may be a very reasonable price for a high quality toy or a wristwatch. These companies want to make their product name a fixture in your home, and their freebies are often surprisingly good in quality. And a few of them are really crap. Read the info about the freebie's size, color and composition carefully. I've seen "duffel bags" the size of a lady's purse.

Many mail-in rebates involve checks for very small amounts of cold, hard cash (like a dollar and a half or two dollars). And this is where mail-ins become exciting and addictive. One woman was recently arrested for obtaining over $30,000 in "fraudulent" rebates. Her basic mistake was that she didn't know "when to say when." Yes, greed is good, but stupidity will stop your personal evolution dead in its tracks. It must have seemed pretty suspicious to the mail-in companies when the same one hundred consecutively numbered apartments would answer their offer in the same month, each bearing remarkably similar grocery store receipts. The most "shocking" part of the whole thing, according to the newspapers, was that the woman in question was obtaining the majority of her proofs of purchase from the trash. And, mind you, this woman who scammed all this money was no genius — that's clear from her mail set-up.

If I wasn't so devoted to pursuing my wealth by other means — means that are more satisfying to me — I would be attempting the same thing as this lady. Without the jail part, of course. And remember this: coupons are an excellent barter item. You just have to find the "coupon clippers" in your area. There are also shady characters who buy coupons *by the pound.* These coupons are used in redemption fraud by stores controlled by organized crime. Some pay as much as $5 a pound. Coupons are, indeed, just like money. (Especially Federal Reserve Notes.)

Perfume Samples

We've covered a lot of things in this chapter which are actually on the "edge" of pure "info diving." No matter. Let's end this chapter with a sweet smell and a good tip. NEVER buy expensive perfume. Lots of discarded magazines contain perfume "samples." Take the perfume-impregnated paper and rub it on sweaty skin. You can use each sample about twice — maybe three times. Mmmm! That's the sweet smell of discarded wealth.

Remember, a good diver is a thinking diver. Feed your head like you feed your body — from dumpsters!

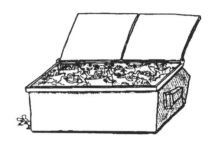

TWISTED IMAGE by Ace Backwords ©1993

Chapter 13
From Novice to Master

In sharing my experiences with you, I almost feel as though I know you, the reader, personally. Together, we have shared the sights, smells, tastes and wealth of the diving lifestyle. I almost feel like we've gone rummaging together.

Books are funny things. You can use a book to increase your knowledge, then go out and apply the things you've learned. But soon you'll encounter something not covered by the book — so you find your own solution. Thus you become wiser than your teacher, the author of the book. The novice becomes the master and outgrows the old master in wisdom. Then a new novice comes along. May it ever be so.

My dumpster diving experiences are not all-inclusive. It seems like every week I learn something new in the trash bins. I've tried to point out the uncharted paths, the unknown places where wealth might be hidden. If you embrace the diving lifestyle and venture on those uncharted paths you will write new chapters. And you will grow wiser than the master, Trashhopper!

I hope this book has been like a garbage container to you — full of unexpected surprises! Wealth! Fun! A book about dumpsters should be a little *like* a dumpster.

Wild Willard Feeds His New Family

You may be wondering, "How did this weird family start doing this sort of thing?" So I'll tell you. I'll also tell you about old rural dumpsites, and diving at the actual city dump.

Dumpster diving began with my dad, a unique individual who did most of his living half a century before I ever met him. After a wild life of boozing, brawling, tramping around the country, jail and soldiering he went from a life on skid row to — incredibly — almost twenty-four years of stability with a wife and three genius children.

I'll give you the highlights and spare you the dates and other dull stuff. He was born on a homestead in Montana, where his first Christmas tree was a tumbleweed decorated with popcorn. He didn't attend school until the age of ten, and then found himself handicapped by the fact he spoke mostly Norwegian. But he became an avid reader, consuming books about every subject. But, he didn't consume those books in school. He would skip school for weeks at a time and spend his days hunting, fishing and trapping — sometimes with an Indian companion.

At 15, he "graduated" from school by completing the 8th grade. He promptly talked his mother and father into falsifying his age so he could join the army. Because of his large size he was allowed to become a machine-gunner and tote the heavy weapon around. At the end of his enlistment — which he spent freezing his ass at Ft. Snelling, Minnesota and assisting with a flood in Arkansas — he joined the army of unemployed workers who rode the rails and looked for jobs in the 1930s. After a few years he joined the army again, negotiating a pretty good deal on the basis of his prior service. The recruiter offered him a choice between the Philippines and Hawaii. Dad flipped a buffalo nickel to decide. It was heads. December 7, 1941 found him a sergeant in the lovely territory of Hawaii, looking forward to a big Sunday breakfast.

The beginning of World War II brought him into the "island hopping" campaign. He and his men sometimes survived on snakes, roots and wild pig while waging war on the Japanese. His island hopping ended at Leyte Gulf when a Japanese grenade shredded sixty pounds of gear on his back as well as the wounded radio operator Dad was carrying to safety. After the war, one of Dad's buddies said they didn't find enough of the radio operator to fill a helmet. Dad awoke in the "dead and dying" section of an emergency field hospital in a Catholic church. A Filipino priest was trying to administer "last rites" to him — a perfectly good Lutheran who happened to carry a "lucky" St. Christopher medal. Thinking the priest was a Jap, Dad grabbed him by the frock and attempted to break his neck. He was unsuccessful, weak from loss of blood, but this action prompted the field hospital to grant him some medical attention.

Dad didn't regain consciousness for long until he found himself on a hospital train in Nevada. He had no identity except the word "sergeant." He was mistakenly listed as "Roman Catholic." His parents had been notified that he was missing in action. Dad had enough pieces of metal in his neck, back, legs and head to keep medical personnel busy cutting out small pieces until the day he died of lung cancer.

He returned home upon medical discharge. He felt like everyone was staring at him because he did weird things like hit the deck when an old lady dropped a hymnal in church. Finally, he just left home and started riding the rails like he had done in the 1930s. He worked as a craps dealer... dock worker... construction worker... lumberjack... you name it. He would work a few months then quit or get fired. For a few years he joined the Air Force and worked in bookkeeping. He went to chef school on his GI Bill. He attempted to join the Marines when Korea flared up, but was medically discharged when he required a chest x-ray. The docs couldn't help but notice that his body was full of cheap metal in hard-to-get-at places.

Somewhere along the way he had two failed marriages. He did time in prison for check fraud. He spent years and years on "skid row" abusing his liver and practicing bad mental hygiene. And dumpster diving.

For several months he worked in a cafe in Billings, Montana. The owner, "Mrs. Kalina," took a liking to him.

"You should meet my daughter, Vernie," she said. "You two would really get along. She runs a little farm that my husband left us."

"Nah," Dad said. "Thanks anyway."

Back to riding the rails and liver abuse. One day he went in a restaurant for a meal. A nice waitress put his dime tip he gave her in the jukebox. She pushed B-1, "Pearly Shells."

"That's my favorite song!" Dad said. "It always makes me think of Hawaii." "You were in Hawaii?" she asked.

Mom had never been further than Montana. They started talking and Mom offered to give him a ride to "Hoffmanville" and set him up with a job at another diner. Along the way, Dad remarked that he knew a Mrs. Kalina in Montana. A relative, perhaps?

A few months later they were married. Everyone — including Dad's parents and the pastor who performed the ceremony — predicted disaster and heartbreak. Mom was in her late 30s and had never married. She went to church three times a week and was a teetotaler. Dad was divorced, an ex-con and a heavy drinker. Friends warned Mom he wanted to murder her and take the farm.

Instead, Mom and Dad were happy as could be. Dad would fall off the wagon from time to time and end up shooting at planes with a deer rifle, but most of the time he managed to remain sober for months or even years at a time. And he never hit any planes with the deer rifle.

Everything wasn't peachy, however. The diner where Mom worked burned down, and she couldn't obtain another job because of her advanced pregnancy. Her small herd of milk cows became sick and died. Dad was "laid off" for no particular reason except he was the "last hired." Even Dad's attempts to poach a skinny deer in the middle of winter were unsuccessful.

So did he whine about his situation? Did he give up? Of course not. By God, after you've been left for dead the universe takes on a sharp-edged clarity for the rest of your life.

"I've been in worse shit than *this!*" he told Mom. "I'll get us some food."

Dumpster diving was old hat to him. He had been scavenging this way for years. However, he had never done it to feed a family. And he had never used a vehicle to haul away his finds.

The dumpster deities looked kindly upon Wild Willard's efforts. He found all kinds of bread, bakery goods, frozen foods and dairy items. The milk and cheese were especially welcome in light of the dead dairy cows. Dad also filled the back of the truck with wooden produce crates. He had often burned these to keep warm in the "hobo jungles." He realized this easy-to-acquire kindling would save him time and effort.

Boldly rummaging behind every grocery store in sight, Dad couldn't believe his good luck. Before, he had sought enough food to feed himself for a day or so, and had to walk from dumpster to dumpster. But there was *so much!* Why, even non-vagrants might find this sort of activity profitable. Why did only vagrants (like he used to be) take advantage of all this free grub?

When he returned home, Dad came into the house holding a box with a dozen quarts of milk. He was as happy and proud as a little kid holding a stringer of fish.

"I *knew* I could find something!" he said. "But it might be sour."

"Doesn't matter," Mom said. "I'll use it. The baby needs calcium."

So you might say I was a dumpster diver from the womb. It's in my bones, so to speak. And everything you've read in this book grew from that small, brave effort to feed a family. Over the years, Mom wrote to her mother in Billings and told her about all the good stuff in dumpsters. So I'm a third generation dumpster diver — white trash, you might say.

Hidden Wealth At Old Rural Dumpsites

Despite my kind words about self-sufficient country folks in the good ol' days, they didn't use *everything*. They threw quite a few things away — like old cast iron toys, tin spice containers, bottles, beer cans. Amazingly, some of this stuff has become valuable — extremely valuable!

Plenty of people prospect around old farms with metal detectors, hoping to find a buried jar of wheat pennies or a lost mercury dime. When their expensive metal detector starts turning up rusted cans, these weekend treasure seekers say, "Oh, hell — this is an old dump!"

Hey, I love to find silver dollars, too. But there's real treasure in that old trash. I've found valuable old perfume bottles, cast iron toys, rare beer cans, old 5¢ cola bottles, and lots of 1930s license plates. Toys are particularly valuable — just *one* can be worth a hundred bucks. Regional beer cans which are no longer produced mean lots to collectors. Some of this stuff might be no older than ten years, but it's still valuable.

A few feet down the discarded items are amazingly well-preserved. But excavating this stuff is hard work and promises no certain reward. You have to love the feeling of finding an old horseshoe, cartridge shell, or a broken "Depression glass" piece. Farmers in my area of Minnesota used to dump whole hayracks full of field rocks in the same places they discarded good junk. You might literally move a ton of rocks to find a rusted tool. Of course, one heavy metallic rock I moved turned out to be a meteorite. And if you find just one old cast iron toy — well, that $50 to $100 (or more!) will soothe your aching muscles a lot.

Personally, I hate lifting weights. I always say to myself, "What's the point? I'm not accomplishing anything but building up my muscles a small bit." But I'll move rocks all day looking for hidden wealth. It gives me a purpose to really work out.

Don't smash that old bottle before you find it! Use sturdy leather gloves to protect your hands and keep your eyes peeled.

Diving At The Dump Itself

When I say "the dump," I'm not talking about a sanitary landfill. You don't need no stinkin' sanitary landfill. Everything in there is squashed daily by bulldozers and covered by a layer of dirt. Some of the poorly-run landfills let things pile up for a few days, and these places are worth scavenging. But the best scavenge is the old-fashioned "town dump." This sacred institution is going the way of the woolly mammoth and nobody gives a damn about preserving old dumps for our children and grandchildren.

It pays to look into your local dump or landfill and find out their set-up. Dumps fall into two important categories: places where picking is allowed and places where it isn't. (Though I hasten to mention ALL dumps are picked. But picking may not be available to the public at large. Watch the bastards mention some shit about "liability.")

Some dumps charge a fee, and scavengers at these dumps look for the most readily salable materials — like copper and aluminum. In the Third World, people scavenge all day to obtain things like paper to sell for a few cents.

Always check the "exchange." This is the place where people leave items for giveaway to scavengers.

Those places where picking *is* allowed have fierce competition. The other humans are bad enough, especially along the U.S.-Mexico border and near Indian reservations. But you also have to compete with dogs, cats, raccoons, skunks, rats, mice and gulls as big as turkeys. In some places there are more exotic pests like wolves, coyotes, vultures, egrets, bears and even bald eagles. Pack a picnic lunch and bring a camera for those special moments.

Of course, wear thick boots (not shoes — boots! THICK boots), thick gloves, disposable clothing and maybe even a mask. I like the cheap masks made for painters which cost about 30¢. Of course, I have a nice supply of top-of-the-line surgical masks from my hospital workplace. A clean rag or handkerchief works well, too.

As dumps burn, they give off noxious fumes like dioxins. Prolonged exposure to dioxins has been implicated in chloracne, a disfigurement of the face, head and neck. The stuff can take *two years to clear up.* If dermatology fascinates you, you can see plenty of chloracne in the cardboard and tarpaper *colonias* of the U.S.-Mexico border. So don't hang around the dump when it's burning, and protect yourself from lingering dioxins by wearing a mask. Of course, industry is a worse source of dioxin pollution than burning dumps, and one just has to wonder if industry is contributing to chronic acne problems.

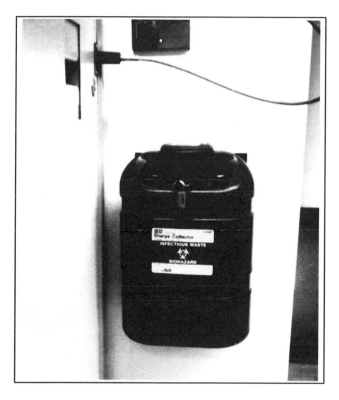

A red "sharps" collector from a hospital.
Stay away from these "biohazards,"
as well as red "hopital waste" bags.

Watch out in those areas of the dump that smell "scorched." You can casually pull a bag aside and have a smoldering area "flare up." People picking trash dumps have died or been severely disfigured this way. Don't smoke. Methane seeps out of the ground around dumps.

Of course, I've mentioned a bunch of unpleasant things for safety's sake. But picking the dump is FUN! You don't have to fret about being observed or caught, you can chat with your partner and take your sweet time. A good dump can produce a truckload of great finds — on an outstanding day you can do better than you would diving dumpsters. If you live near a dump take advantage of your luck.

The big problem with dumps is the lack of selectivity. Stuff from apartments, barber shops, car dealerships and grocery stores is all mixed together. The "crusher" on the garbage truck does a lot of damage, but not as much as most people assume. A couch will be destroyed beyond use, for example, but stuff in cans usually arrives intact. Use a stick to pick through things and minimize your exposure to bad smells and stuff that will adhere to your clothing. Bag blades are essential equipment. I've found that a scythe with a long handle can save you a lot of bending over. But keep it sharp! Be careful walking around on trash bags — don't fall down and slice yourself.

Dumpsters are filled with little stories, but a dump is like a big, sprawling novel about the daily life of the city. Scavengers find bodies and body parts more frequently than the papers reflect. I've done a lot of scavenging and never found a human being — but I was a hundred yards away when Slash found a human fetus.

He put it in a shoe box and took it home. Thus began a weird relationship that lasted about a year. After cleaning it up, he put it in a deluxe specimen jar filled with formaldehyde. He used two clear plastic blocks to make it stay upright, rather than lying on its back.

The fetus looked delightful floating in serene, preserved solitude. Slash would talk to her. And he gave her a name — Salina. In fact, he developed what struck me as a rather unhealthy attachment to her.

Eventually, however, a friend of Mrs. Spooner's, the head of the Minnesota chapter of the National Christian Pro-Life Council came from the Twin Cities and offered to buy Salina. He wanted to use her in an anti-abortion display, to show how human-like a fetus looks.

Slash — very, very reluctantly — said "Goodbye" to Salina, declining to accept any payment for her.

The Dumpster Is NOT The End

I hoped to God Salina wouldn't come home like Lassie did. It still creeps me out to talk about it. To think we found something like that in the trash!

The dumpster is NOT the end... it's the beginning. What mysteries await on the other side of the lid? Wealth? Survival? Fun? Horror? All of the preceding?

You won't know unless you take the leap into the unknown. I believe riches await you.

Remember — THAR'S GOLD IN THEM THAR DUMPSTERS!

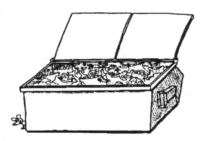

TWISTED IMAGE by Ace Backwords ©1993

Chapter 14
Trashy Treats

Cooking with dumpster food is no different than regular cooking except you must pay particular attention to cooking everything well. (To avoid bacteria.) The more skillful you are, the easier you will find it to "whip something together." But these treasured recipes from the Hoffman family may be especially helpful to you. *Bon apetit!*

Bad Banana Whipped Cream Substitute

Add a soft, overly ripe banana to stiffly beaten egg whites and whip until fluffy. Add a drop or two of vanilla or a few spoonfuls of melted vanilla ice cream. It doesn't taste like whipped cream, but it's damned good topping. You might try it with:

Fruit Cobbler Ala "Slash"

Jed can't get enough good fruit cobbler with bad banana whipped cream substitute.

Peel the fruit, removing only the very worst spots. Don't worry about soft spots. Cut into chunks and spread 3 to 4 cups of fruit on the bottom of a buttered baking dish. Add a dash of cinnamon and dot the fruit with butter or margarine. Mix up a simple biscuit dough or use discarded pizzeria dough, bread dough, frozen ready-to-bake breakfast rolls, or whatever. Spread over the fruit and bake at 350º for half an hour. Mmm!

Stale French Toast

Stale bread is better than fresh bread for French toast. Soak the bread in a saucer of milk (even sour milk) until soft — not spongy. Dip each piece into a beaten egg and fry in butter. Sprinkle with cinnamon before serving.

Bad Eggplant Caviar Substitute

Cook the eggplant in boiling water until it is tender. Let cool, peel, chop fine. Sauté chopped eggplant in 2 to 3 teaspoons of olive oil with ½ cup chopped onion, one small peeled and chopped tomato, and two teaspoons lemon juice. (You can obtain juice from an old, dried out lemon by boiling it.) When most of the liquid has evaporated from the mixture, salt and pepper to taste. Chill before serving ice cold with dumpster dived crackers. Does it taste like caviar? Alas, no. But it's quite good.

Banana Soul Bread

It's called "soul bread" because you use black bananas.

Use three overripe bananas, mashed. ½ cup butter or margarine. 1 cup sugar. Two eggs. 1 teaspoon soda and 2 cups flour. Mix all ingredients in a well-greased pan. Bake for 45 minutes (approx.) at 300º.

Stale Peanut Butter

Place 2/3 cup stale peanuts plus 2 teaspoons of corn, peanut or other oil in a blender. ½ teaspoon of salt optional. You can't tell the difference between stale and fresh peanut butter.

Candied Citrus Peel

Waste not, want not. Cut citrus peels into narrow strips. Put two cups of peels in a pan with two cups of water and simmer ten minutes. Drain, add more water, simmer again. Repeat twice more. Now make a thick syrup from ½ cup water and 1 cup sugar. Add the peel to the syrup and boil until the liquid is absorbed. The peel turns translucent but retains its bright color. Spread on a rack to dry. Roll in sugar, chocolate, or eat plain. These cost a *lot* in specialty shops.

Pea Pod Soup

Put pea pods in a linen bag and boil to extract the flavor. Makes a light, refreshing soup. Add carrots, onions or what's available.

Dumpster Burger Casserole

You will need two pounds of hamburger patties from your local fast food dumpster. Leave the ketchup, pickle and cheese on the patties. Don't worry about a little bread stuck to the patties, either. You will also need three cups of cooked elbow macaroni (cheap!) and one onion. Use your favorite spices, such as garlic salt, pepper, Mrs. Dash, etc. You'll also need two cans

of tomato sauce (cheap!) or a can of tomato soup. Use whole tomatoes (soft ones are fine) or a can of cheap veggies like green beans, peas, corn, etc. Use both, if you like. Remember to drain the cans.

Chop up the hamburger patties and heat briefly. (They're already cooked.) Mix cooked macaroni, tomato sauce or tomato soup, spices and veggies with the hamburger. Top with cheese if desired and available. (You can use the cheese from the burgers instead of mixing it into the casserole if you want the thing to look pretty.) Bake at 350º for 45 minutes. Use the hamburger buns to make garlic toast.

Enjoy, enjoy.

Cow Manure To Go

Can't find a better place for this last "recipe," which is a clever and convenient way to obtain fertilizer without driving to the country and trying to buy some cow-pies.

Obtain plastic trash bags full of leaves and throw a gallon of water in the bags. Seal it up and let the bacteria work. Leave in the warm sun for best results, but do not allow to freeze. Within 20 to 25 days (depending on temperature, leaf composition, etc.), you won't have compost, but *manure*. And it's packaged and ready to use!

Tie the bags loosely to let gas escape. Do this away from buildings, as you may note a sour "silage" smell.

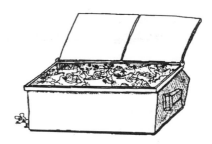

YOU WILL ALSO WANT TO READ:

☐ **14116 Building With Junk And Other Good Stuff,** *by Jim Broadstreet.* A complete guide to building and remodeling using recycled materials. Millions of dollars worth of building materials are thrown away every day. This book shows how to find, store and use this good stuff. Covers floors, ceilings, walls, foundations, roofs, plumbing, wiring, utilities, windows. *1990, 8½ x 11, 159 pp, illustrated, hard cover.* **$19.95.**

☐ **13044 Guerrilla Capitalism, How to Practice Free Enterprise in an Unfree Economy,** *by Adam Cash.* What good is "believing in" free enterprise if you don't practice it? This book gives you step-by-step instructions on how to do business "off the books;" doing business without a license; getting customers to pay in cash; keeping two sets of books; investing unreported income; and much more. Highlighted with case histories of successful guerrilla capitalists. *1984, 5½ x 8½, 172 pp, illustrated, soft cover.* **$14.95.**

☐ **13077 How To Make Cash Money Selling At Swap Meets, Flea Markets, Etc.,** *by Jordan Cooper.* After years of making good money at flea markets, the author lets you in on the secrets of success. What to Sell; Transportation; Setting-Up; How to Display Your Wares; Pricing; Bad Weather; The IRS; and much more. *1988, 5½ x 8½, 180 pp, illustrated, soft cover.* **$14.95.**

☐ **85120 Twisted Image,** *by Ace Backwords.* This is the first collection of comic strips by America's funniest underground cartoonist. Ace Backwords takes on the controversial topics of sex, drugs and modern culture. His strips have appeared in more than 200 "marginal" publications including High Times, Maximum Rock 'n' Roll, Screw and the Loompanics Catalog. For adults only. *1990, 8½ x 11, 128 pp, more than 200 strips, soft cover.* **$12.95.**

☐ **40079 How To Steal Food From The Supermarket,** *by J. Andrew Anderson.* Written by a supermarket security guard, this book will give your budget a boost! Learn all the ins and outs of shoplifting success, including, ● Do-it-yourself markdowns ● Scamming the scanner ● How to dress for success ● Defeating store security ● And much more, including the one mistake that trips-up most shoplifters and the one item you *must* bring shoplifting with you. *This offer not available in stores. 1993, 5½ x 8½, 63 pp, soft cover.* **$10.00.**

☐ **17056 Freedom Road,** *by Harold Hough.* Have you dreamed about leaving the rat race but don't know where to start? This book will take you down the road to freedom, one step at a time. It will show you how to make a plan, eliminate your debts, and buy an RV. You'll learn about beautiful places where you can live for free. You'll learn how to make all the money you need from your hobbies. And you'll learn how to live a comfortable, healthy lifestyle on just a few dollars a day. Why wait for retirement when you can live a low-cost, high travel lifestyle today? Get on Freedom Road! *1991, 5½ x 8½, 174 pp, illustrated, soft cover.* **$16.95.**

And much more. Please see our catalog ad on the next page.

- -

Loompanics Unlimited/ PO Box 1197/ Port Townsend, WA 98368

Please send me the titles I have checked above. I have enclosed $ _____ (which includes $4.00 for shipping and handling of 1 to 3 titles, $6.00 for 4 or more).

Name _____

Address _____

City/State/Zip _____

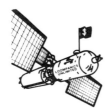